Sustainable Development Goals Series

World leaders adopted Sustainable Development Goals (SDGs) as part of the 2030 Agenda for Sustainable Development. Providing in-depth knowledge, this series fosters comprehensive research on these global targets to end poverty, fight inequality and injustice, and tackle climate change.

The sustainability of our planet is currently a major concern for the global community and has been a central theme for a number of major global initiatives in recent years. Perceiving a dire need for concrete benchmarks toward sustainable development, the United Nations and world leaders formulated the targets that make up the seventeen goals. The SDGs call for action by all countries to promote prosperity while protecting Earth and its life support systems. This series on the Sustainable Development Goals aims to provide a comprehensive platform for scientific, teaching and research communities working on various global issues in the field of geography, earth sciences, environmental science, social sciences, engineering, policy, planning, and human geosciences in order to contribute knowledge towards achieving the current 17 Sustainable Development Goals.

This Series is organized into eighteen subseries: one based around each of the seventeen Sustainable Development Goals, and an eighteenth subseries, "Connecting the Goals," which serves as a home for volumes addressing multiple goals or studying the SDGs as a whole. Each subseries is guided by an expert Subseries Advisor.

Contributions are welcome from scientists, policy makers and researchers working in fields related to any of the SDGs. If you are interested in contributing to the series, please contact the Publisher: Zachary Romano [Zachary.Romano@springer.com].

More information about this series at http://www.springer.com/series/15486

David B. Abraham • Seema D. Iyer
Editors

Promoting the Sustainable Development Goals in North American Cities

Case Studies & Best Practices in the Science of Sustainability Indicators

Springer

Editors
David B. Abraham
Department of Kinesiology
Rice University
Houston, TX, USA

Seema D. Iyer
Department of Finance and Economics
University of Baltimore
Baltimore, MD, USA

ISSN 2523-3084 ISSN 2523-3092 (electronic)
Sustainable Development Goals Series
ISBN 978-3-030-59172-4 ISBN 978-3-030-59173-1 (eBook)
https://doi.org/10.1007/978-3-030-59173-1

This Springer imprint is published by the registered company Springer Nature Switzerland AG
The registered company address is: Gewerbestrasse 11, 6330 Cham, Switzerland

Contents

1 **Introduction: Localizing SDGs and Empowering Cities and Communities in North America for Sustainability** 1
David B. Abraham and Seema D. Iyer

2 **Localizing the SDGs in Baltimore: Challenges and Opportunities of the USA Sustainable Cities Initiative** 7
Seema D. Iyer

3 **Los Angeles: Measuring Its Ambition to Achieve the SDGs** 21
Tony Pipa

4 **The Idea of Sustainability to Evaluate Growth and Development in the Houston Region** 31
David B. Abraham

5 **Making the SDGs Relevant for Cities: Using the Community Capital Tool in British Columbia** 51
Maria Spiliotopoulou and Mark Roseland

6 **How Local Communities Can Align with United Nations' Sustainable Development Goals: A Santa Cruz County Case Study** 67
Susan Brutschy, Jennifer Anderson-Ochoa, James Connery, and Donna R. Maurillo

7 **Top-Down and Bottom-Up Approaches to the SDG Monitoring Challenge** 87
Jessica Espey

8 **Data for Good for All: Enabling All Communities to Track Progress Toward SDG Implementation** 97
Jennifer Temmer and Stefan Jungcurt

9 **Helping the Neighborhood: Creating a Sustainability Indicator of Substandard Housing** 109
Luis Estevez

10 **How the Youth Social Entrepreneurship (YSE) Model Supports Implementation of SDG #8 to bring Economic Growth and Decent Work in American Cities** 127
Peg Thomas, Nancy Jacobs, Jennifer Valorose, and Laura Schauben

11 **Local SDG Indicators for US Cities and Communities** 147
David B. Abraham

Index 157

Contributors

David B. Abraham Rice University, Houston, TX, USA

Jennifer Anderson-Ochoa Applied Survey Research, Watsonville, CA, USA

Susan Brutschy Applied Survey Research, Watsonville, CA, USA

James Connery Applied Survey Research, Watsonville, CA, USA

Jessica Espey Sustainable Development Solutions Network, New York, NY, USA

Luis Estevez St. Cloud State University, St. Cloud, MN, USA

Seema D. Iyer University of Baltimore, Baltimore, MD, USA

Nancy Jacobs Sundance Family Foundation, St. Paul, MN, USA

Stefan Jungcurt International Institute for Sustainable Development, Gatineau, QC, Canada

Donna R. Maurillo Applied Survey Research, Watsonville, CA, USA

Tony Pipa Brookings Institution, Washington, DC, USA

Mark Roseland Arizona State University, Tempe, AZ, USA

Laura Schauben Wilder Research, St. Paul, MN, USA

Maria Spiliotopoulou Simon Fraser University, Vancouver, BC, Canada

Jennifer Temmer International Institute for Sustainable Development, Simcoe, ON, Canada

Peg Thomas Sundance Family Foundation, St. Paul, MN, USA

Jennifer Valorose Wilder Research, St. Paul, MN, USA

Introduction: Localizing SDGs and Empowering Cities and Communities in North America for Sustainability

1

David B. Abraham and Seema D. Iyer

1.1 The Sustainable Development Goals

In September 2015, the member countries of the United Nations adopted the Sustainable Development Goals (SDGs), which defined new standards for a global commitment to the three interrelated pillars/objectives: economic development, social development, and environmental development. The SDGs form a cohesive and integrated package of global aspirations framed as Goals that individual countries commit to achieve by 2030. The 17 SDGs address the most pressing global challenges of our time, calling upon collaborative partnerships across and between countries to address universal, integrated challenges to sustainable development. The SDGs include goals for addressing job loss, deteriorating infrastructure, social exclusion, and climate change, among many other issues facing societies today. The SDG agenda comes at a time when more than half of the world's population lives in urban areas. As this portion of the population grows at a rapid speed, so too do the complex development challenges in these locations.

The SDGs can provide a long-term and sustainable approach to city planning by providing a suite of clear, common, and objective Goals that can be continuously pursued irrespective of political cycles. Ensuring full ownership of the goals through an inclusive, participatory dialogue is of paramount importance to the success of the SDG agenda. The goals must ultimately act as the common language for government, business, and citizens and represent a shared ideal to be pursued at the city level but also within each community. The SDGs, therefore, must be localized through a fully representative and inclusive participatory planning process.

1.2 Overview of the Sustainable Development Goals

The SDGs provide a comprehensive and inclusive framework for sustainable development that can be utilized by local stakeholders as a guide for community planning. The history and main principles of the SDGs are outlined below.

"Getting Started with the SDGs in Cities: A Guide for Stakeholders[1]" summarizes the SDGs as follows:

D. B. Abraham
Rice University, Houston, TX, USA
e-mail: DAbraham@Rice.edu

S. D. Iyer (✉)
University of Baltimore, Baltimore, MD, USA
e-mail: siyer@ubalt.edu

[1]Prepared by the Sustainable Development Solutions Network, 2016.

D. B. Abraham, S. D. Iyer (eds.), *Promoting the Sustainable Development Goals in North American Cities*, Sustainable Development Goals Series, https://doi.org/10.1007/978-3-030-59173-1_1

> The Sustainable Development Goals (SDGs) are a universal set of goals, targets and indicators[2] that UN member states have committed to use, to frame both domestic and international development policies over the [...period 2015–2030]. They build upon the progress of the Millennium Development Goals (MDGs), which were agreed by governments in 2001 and expired in 2015. While the MDGs focused on reducing extreme poverty in all its forms, the SDGs pursue a broader agenda that encompasses the social, environmental and economic aspects of sustainable development, which is relevant for all countries worldwide.
> The SDGs are at the core of the 2030 Agenda for Sustainable Development,[3] which was ratified by all UN member states at the 2015 United Nations General Assembly. Their 17 goals and 169 targets address critical issues facing the world today, including the eradication of extreme poverty, tackling global inequality and climate change, promoting sustainable urbanization and industrial development, protecting natural ecosystems, and fostering the growth of peaceful and inclusive communities and governing institutions (p. 8).

The SDG framework addresses five key themes, referred to as the Five Ps of Sustainable Development:

People – The commitment to ending extreme poverty, hunger, and economic and gender inequality

Planet – The commitment to protect the planet from degradation through sustainable development, production, consumption, and natural resources management practices and to address the causes and effects of climate change

Prosperity – The adoption of consumption and production patterns that are sustainable for future generations and result in equitable economic growth and participation for all members of society

Peace – The promotion of good governance, rule of law, anti-corruption, human rights, and equal protection under the law for all members of society

Partnerships – The coordination of a multitude of stakeholders, including national and local governments, multinational corporations, NGOs, and other members of global civil society to implement the SDG agenda with accountability and transparency[4]

The SDG framework is also designed to reflect advancements in the field of development since the advent of the MDGs, as it promotes key opportunity areas for improved outcomes by applying these five principles in development strategies:

Inclusivity – The SDG agenda stands on the principle that *no one is to be left* behind and therefore requires the engagement of stakeholders across all levels of society in order to effectively account for and respond to the needs and interests of all.

Universality – In order to achieve global targets for development, the involvement of developed and developing countries is required. Understanding that development contexts vary worldwide, the SDGs are designed to be adapted as they are applied to local situations.

Integration – The SDG agenda addresses the complexity of long-term solutions, by recognizing the interconnectivity of development policies and investments and building on existing relationships between stakeholders across the three dimensions of sustainable development: economy, environment, and society.

Technologically driven – Advances in ITC and data availability inform sustainable development policy and investment as they improve global communication and interconnectedness and bring to light a range of data that illustrate and measure development needs, challenges, and progress.

[2]A complete list of the SDGs and their targets is available at https://sustainabledevelopment.un.org/sdgs

[3]United Nations Resolution A/RES/70/1 of 25 September 2015 (2).

[4]For more information on the "5 Ps," see: SDSN, "Getting Started with the Sustainable Development Goals – A Guide for Stakeholders", (New York: SDSN, 2015), 8–9. Available at: http://unsdsn.org/wp-content/uploads/2015/12/151211-getting-started-guide-FINAL-PDF-.pdf

Locally focused – Local achievements in sustainable development and the SDGs require support, action, and coordination from communities and local governments. In this respect, cities are critical centers of sustainable change due to their population density and economic needs and output. For this reason, piloting the agenda in cities like Baltimore will not only provide benefits to the community, but the bottom-up approach can also inform national-level policy, and it will serve as an example for other cities around the world that seek ways to participate in the SDG initiative.[5]

1.3 Why the SDGs Are Useful for Cities

"Cities are where the battle for sustainable development will be won or lost".[6] This is a salient point, as the SDGs have come into effect in a world that is increasingly urban. A little over half the global population currently resides in cities, and by 2050 this statistic is expected to grow to two-thirds of the global population.[7] Urbanization has created some of the world's most complex development challenges. This trend of urbanization, however, also yields the opportunity to create high-impact solutions, as a result of the agglomeration of people and business activity. The imperative for innovation and change is spurring wide interest and investment in twenty-first-century urban development.

Mayors and local leaders who are tasked with the responsibility to manage and improve the quality of life in urban environments recognize that the SDGs provide a road map for balanced and equitable urban development.[8] Additionally, the quest to build sustainable cities that advance global progress is putting mayors and local government leaders at the forefront of change. This trajectory is triggering broad interest and investment in urban development. The SDGs provide a set of integrated objectives that comprise a more complete and sustainable vision of urban development, which provides equal living and working opportunities to all inhabitants, to promote healthy living environments and resilience against the array of everyday challenges and risks that we face today. As UN Secretary General Ban Ki Moon described it, the SDGs are a "people's agenda." Utilizing the SDG framework therefore helps to show responsivity to what local residents want, the world over.

1.4 How This Volume Is Organized

Chapters 1, 2 and 3 of the book describes the uptake and performance of the SDGs in three US cities. Baltimore had been among the first cities to attempt the exercise at a time when very little was known about the global Goals among local leadership and multi-sector civil society stakeholders involved in sustainable development. Dr. Seema Iyer describes how leveraging the expertise of the local community indicators project (Baltimore Neighborhood Indicators Alliance) helped fill in gaps in measures of equity and justice and how the SDGs remained relevant during the city's update to the Sustainability Plan in 2019. In contrast, the next case from Los Angeles demonstrates that even with strong leadership from Mayor Garcetti, alignment with the global goals can only happen alongside internal alignment of agencies, policies, and data. Tony Pipa

[5] "Getting Started with the Sustainable Development Goals – A Guide for Stakeholders," (New York: SDSN, 2016), 10–11. Available at: http://unsdsn.org/wp-content/uploads/2015/12/151211-getting-started-guide-FINAL-PDF-.pdf

[6] High Level Panel of Eminent Persons on the Post-2015 Development Agenda, 2013. http://www.post2015hlp.org/the-report/

[7] UNDESA, "World Urbanization Prospects-The 2014 Revision," United Nations, New York, 2014. Available at: https://esa.un.org/unpd/wup/Publications/Files/WUP2014-Highlights.pdf

[8] "Getting Started with the Sustainable Development Goals – A Guide for Stakeholders," (New York: SDSN, 2016), 10–11. Available at: http://unsdsn.org/wp-content/uploads/2015/12/151211-getting-started-guide-FINAL-PDF-.pdf

shows that as LA prepares for the 2028 Olympics, the SDG framework can help set priorities for ongoing work towards a comprehensive vision for the city. In Houston, Dr. David Abraham reveals the city's history of engagement with sustainable development thought leaders and institutions, dating back from the 1970s to today. However, as Dr. Abraham points out, in his comparison of SDG performance in Houston and the top 100 most populous metros in the United States, Houston stakeholders need to tackle the fundamental issue of sprawl to significantly improve the sustainability performance of the region.

Although the introduction of the SDGs was new in North American cities in 2015, the idea of measuring and tracking sustainable development was certainly not. Chapters 5 and 6 of the book provide city-level examples of *mapping and adapting* existing sustainable development frameworks established in North America to the SDG framework. Maria Spiliotopoulou and Dr. Mark Roseland discuss the urgency for local action for achieving the global goals so that utilizing an adopted framework better mobilizes action toward implementing and monitoring the SDGs locally. They provide research findings for a complex matching and mapping exercise to show the relationship between the SDGs, the Community Capital Framework, and local goals for municipalities in British Columbia. The team of authors from California-based Applied Survey Research describes how and why it was possible to connect their long-standing work evaluating the Community Assessment Project in Santa Cruz County to the global goals. They explain that three conditions are necessary for any framework to successfully apply the SDGs locally: (1) a prior commitment to well-being for all which aligns with the priorities of the SDGs for leaving no one behind, (2) experience with creating a measurement system that tracks well-being, and (3) the ability to coordinate action across multisector stakeholders.

Chapters 7 and 8 of the book are about the dual challenge of both *national mobilization and local engagement* of SDGs. Indices and online Community Indicator System (CIS) portals are recommended, respectively, as national- and local-level tools to manage and track local-level performance differences on SDGs. Jessica Espey of the Sustainable Development Solutions Network (SDSN) writes about the group's opportunities and challenges in preparing a national US Index for the SDGs. She presents the need to also complement such an effort with local-level action on SDGs, since city-level mayors are politically and thematically key stakeholders in developing sustainability policy. Jennifer Temmer and Stefan Jungcurt of the International Institute for Sustainable Development (IISD) focus on the thematic benefits of community and city-level adaptation of the SGDs, before they present the benefits of a novel CIS portal their group developed for use by cities interested in reporting on the SDGs. Their research points to five predominant thematic indicator measurement movements that can be addressed by implementing the SDGs. These are Quality of Life, Healthy Communities, Sustainability, Government Performance and Benchmarking, and "Subjective Well-Being," which incorporates Public Happiness and Life Satisfaction. This shows that the SDGs are functional enough to report performance on many of the contemporary and important movements driving urban areas in North America.

Chapters 9 and 10 of the book are about *new indicator definitions and methods* that groups have developed to address unique development issues of importance in North America. The SDG indicators do not cover the full complexity of all development issues in North America. North American countries and cities are unique in the vast amount of research and resources that have gone into tackling wicked problems of importance. As such, more robust methods are needed to analyze and capture the nuances that may exist within themes. According to the 2017 report of the Secretary-General to the United Nations, "Progress towards the Sustainable Development Goals," during the 10-year period from 2007 to 2016, 89 percent of countries around the world conducted one housing census, while 25 countries did not have such a data source. The papers in this section present examples from Youth Development and Housing. Both examples developed are composite indicators. This reflects the recognition of complexity that can be found within our ability to define and tackle problems in urban development. Specifically, Dr. Luis Estevez presents an academic method for identifying substandard housing within a community. Peg Thomas et al. from the Sundance Family Foundation present a methodology from the professional field. They present a *multi-indicator/multi-structured model* for providing policy guidance around developing more equitable economic opportunities for individual youth and their communities.

Chapter 11 of the book is a proposed heuristic outline of the SDGs localized for North American cities. Dr. Abraham, and his team from Rice University, will use this outline as the basis for a multi-city iterative process in 2020 to garner broad support for an agreed upon set of indicators for North American cities.

1.5 Recurring Themes in This Volume on SDGs for North American Cities

1. *Mapping local priorities, programs, and indicators to the national SDGs*. This is a task that has to be locally performed and communicated. The ability to tailor the localization approach is a hallmark feature of the SDGs.
2. Localization has both been able to *leverage existing local indicator systems* and *learn from the experiential knowledge* these local systems have attained.
3. When localization has occurred, the impetus has *predominantly come from the bottom-up in most of these cities*. Even in the cities that were invited to be a part of the USA-SCI, these pilot projects were only possible with local, on-the-ground groups making the case.
4. *Low awareness of what the SDGs are and lack of leadership in North American cities will hamper future attempts to localize them*. More work needs to be done to bring the growing international excitement to North American cities.
5. *The SDG framework can be used to highlight gaps in local indicator systems or priorities*. The SDGs represent a new language or way to communicate the comprehensive themes in urban- and country-level development.
6. *Local jurisdictions in the United States are strategic for implementing SDGs since they are notable for having broad autonomy* in decision-making and adoption and implementation of policies, as long as these do not contradict state or federal laws or statutes.

2 Localizing the SDGs in Baltimore: Challenges and Opportunities of the USA Sustainable Cities Initiative

Seema D. Iyer

2.1 Introduction

In the same year that the member countries of the United Nations adopted the Sustainable Development Goals (SDGs), the city of Baltimore was in the grips of challenging times. In late April 2015, the death of a young black man, Freddie Gray, while in police custody set in motion civil and racial unrest at a level not seen in Baltimore and many US cities since 1968. In the wake of significant physical and emotional distress in the city, many community-based organizations, foundations, and civic groups, including a newly formed non-profit in response to the unrest called One Baltimore, galvanized into action by bringing people in Baltimore together to reflect on what happened, to help everyone heal from not only the acute trauma of the unrest but also the chronic conditions that led to such an uprising. However, within a few months, then-mayor Stephanie Rawlings-Blake announced that she would not be seeking reelection in 2016, which heralded an unprecedented field of 14 different candidates vying for the Democratic primary nomination and the attention of the voting electorate in Baltimore.

With this backdrop, when the UN Sustainable Development Solutions Network (SDSN) selected Baltimore as one of three cities to participate in the launch of the USA Sustainable Cities Initiative (USA-SCI) in September 2015, it was clear that there would be many challenges to overcome to take advantage of a yet-unknown set of potential benefits. The objective of the USA-SCI program was to bring the global agenda down to the local level of the American urban environment by piloting a process of "localizing" the SDGs in three cities: New York, NY; San Jose, CA; and Baltimore, MD. The approach was conceived to initiate collaborations among academic institutions and non-profit organizations in each of these cities to support the development of city-level development strategies that align with the 17 SDGs.

For Baltimore, the invitation to be a part of USA-SCI represented a moment of opportunity to be at the forefront of a global conversation. However, with such traumatic events stemming from the death of Freddie Gray and a vacuum in local executive leadership, the localization process in Baltimore relied on leveraging and highlighting the strength of local stakeholders involved in sustainable development. Much of what would become the SCI-Baltimore initiative sought to raise awareness about the newly adopted SDGs among the strong civil society sector in Baltimore consisting of non-profits, philanthropy, community-based organizations, and advocacy groups.

S. D. Iyer (✉)
University of Baltimore, Baltimore, MD, USA
e-mail: siyer@ubalt.edu

D. B. Abraham, S. D. Iyer (eds.), *Promoting the Sustainable Development Goals in North American Cities*, Sustainable Development Goals Series, https://doi.org/10.1007/978-3-030-59173-1_2

One of the key partners chosen for the localization effort was the Baltimore Neighborhood Indicators Alliance (BNIA) at the University of Baltimore, which is the local member of the National Neighborhood Indicators Partnership (NNIP). Since 2000, BNIA has served as the "data intermediary" for Baltimore, focusing on acquiring and disseminating local data to neighborhoods and other multi-sector stakeholders. Having the local community indicators project be a part of the SDG localization process became critical to ensuring Baltimore saw the effort to fruition for reasons which will be discussed in this chapter. First, given its longstanding mission to improve quality of life in distressed communities, BNIA already had deep connections with local governmental agencies as well as neighborhoods impacted by the unrest without seeming like an intrusive outsider during a sensitive time in the city. Second, BNIA was able to quickly align existing, locally relevant indicators to the SDG targets and identify gaps in data that allowed stakeholders to focus on ways to measure equity and justice.

Over the course of a year the USA-SCI effort in Baltimore yielded a wealth of insights and ideas for furthering inclusive, coordinated sustainable development efforts in Baltimore. This chapter provides a glimpse of how the process of localization unfolded in Baltimore and how the challenges and opportunities became clear over time which serves as a resource for local stakeholders in other US cities, providing a summary of current achievements of SCI-Baltimore and recommendations for achieving long-term, equitable sustainable development benefits for Baltimoreans by aligning local efforts with the SDGs.

2.2 Baltimore and the USA Sustainable Cities Initiative

As one of the oldest cities in the USA, Baltimore is a vibrant and diverse community that nevertheless faces significant development challenges, such as depopulation, inequality, poverty, unemployment, and infrastructure degradation. For example, in 2017, 32.9% of children in Baltimore lived below the poverty line, in comparison to the national average of 20.3% (American Community Survey). Similar to a number of other cities, revitalization tends to be concentrated around the downtown area and, in the case of Baltimore, it includes the scenic waterfront of the city's historic harbor. High-rise office buildings in the historic downtown have been converted into luxury living spaces for millennials and wealthy empty-nesters. A short distance away from the city center, however, are over 16,000 vacant houses along with significant abandoned industrial sites and strikingly empty storefronts. As American manufacturing declined, so went the jobs in Baltimore as the city continues to adapt to the new, post-industrial age.

In 2012, the Department of Public Works reported the level of lead in Baltimore's drinking water at the Environmental Protection Agency "action level" of 15 ppb, indicating the water unsafe for children and pregnant women to consume through drinking and cooking. Furthermore, Baltimore Harbor continues to experience unhealthy levels of pollution due to the city's beleaguered sewer system, which is causing damage to the natural ecosystem and restricting people's access to the water.

Beginning in September 2015, the SCI-Baltimore team (described below) worked to develop the institutional infrastructure for engaging multi-sector stakeholders who could provide substantive input into the establishment of quantitative targets and indicators as part of an ongoing effort to localize the SDGs and integrate their comprehensive principles into the city's development activities. The primary activities of the process, described in this chapter, were (1) convening local stakeholders, (2) stocktaking of existing plans and policies, and (3) choosing indicators for tracking the SDGs in Baltimore.

2.3 Convening Baltimore Stakeholders

2.3.1 The Core Project Team

The University of Baltimore (UB), which had participated previously in the regional consultations[1] held during the development of the SDGs, was selected by SDSN as the "host" for the SCI-Baltimore program.[2] The College of Public Affairs and the Baltimore Neighborhood Indicators Alliance (BNIA) provided resources and expertise to the initiative by planning and serving as lead convener for a series of consultative discussions that aimed to (a) build and contextualize knowledge of the SDGs, (b) inform a stocktaking exercise of current development plans and actors, and (c) develop and refine a set of proposed indicators for measuring development achievements aligning with the SDGs. The entire project team leveraged existing relationships and the community in Baltimore to convene key stakeholders in the service of the initiative's goals and to operationalize credible indicators that reflect community concerns. In addition to the resources of the University of Baltimore, the core team included two additional partners: University of Maryland (UMD) and Communities Without Boundaries International (CWBI). UMD's National Center for Smart Growth played an integral role in the review of potential SDG indicators for Baltimore context and supported outreach to stakeholders focused on environmental sustainability activities. CWBI, a nongovernmental organization that supports community dialogue in locations worldwide, augmented the discussions convened by UB by hosting meetings for community leaders to provide feedback on SCI-Baltimore activities, to reflect on technical discussions about targets indicators, and to consider their roles in SDG implementation.

[1] United Nations Association of the USA (2014). Recap of "Maryland Inter-Generational Consultation on UN Development Goals" http://www.unanca.org/news-events/news/363-recap-of-q-maryland-inter-generational-consultation-on-un-development-goals

[2] In July 2014 the city of Baltimore designated the university as one of eight "Anchor Institutions" that provide vital resources and support for the city's sustainable growth. UB was chosen as an Anchor Institution for its history of providing innovative and accessible education to a diverse population in an urban setting and for its positive economic impact including over $275 million in direct investment. The University of Baltimore fosters research and education in highly relevant areas such as public policy evaluation and global affairs and houses research centers including the Baltimore Neighborhood Indicators Alliance.

2.3.2 SCI-Baltimore SDG Executive Team and Working Groups

Over the course of the year, several events were convened to consult local organizations, experts, and authorities on the stocktaking exercise and indicator development and to establish a community of practice to coordinate numerous local sustainable development activities and to promote SDG achievement over the long term. The Baltimore SDGs Executive Team (SDG-ET) brought together representatives from key organizations who the project team knew had knowledge of sustainable development strategies and sustainable development data for Baltimore in order to advise on the overall SCI-Baltimore process, review the relevance and appropriateness of SDG targets and indicators that were revealed by the preliminary stocktaking exercise, and brainstorm additional organizations and initiatives to include in the SCI-Baltimore effort. Consequently, an expanded list of "SDG partners" were convened as working groups to discuss proposed SDG indicators for Baltimore that could be used to set appropriate and realistic targets and track them over time. In each of these meetings, participants were familiarized with the SDGs, the stocktaking exercise on existing plans that address these goals in Baltimore/Maryland, and proposed indicators that can be measured and tracked annually to benchmark current conditions in Baltimore.SDG partners participated in the working groups in the area of "People," "Prosperity," and "Planet," as defined by the

SDG's 5 P approach (including Peace and Partnerships) in order to apply their technical expertise. The working groups then came together in plenary discussions to integrate their perspectives toward a set of indicators that would help Baltimore track progress toward the global targets. SDG partners were also provided ways to promote the SDGs within their networks. This included:

1. Attending any of the convenings of the SCI-Baltimore process and use #SDGBaltimore to broadcast via social media how those discussions related to the SDGs.
2. Taking the *"Which Goals Are You?"* Quiz[3] to help working group members personalize their understanding of SDG priorities and then using this information in organizational discussions, community consultations, and other public events.
3. Informing the project team and the SDG-ET of additional community-based forums that working group members could attend or should be present at to further raise awareness about the SDGs and the SCI-Baltimore process.

2.3.3 Expanding Inclusion Through "Listening-to-the-Listening"

For sustainable development efforts in Baltimore to be truly inclusive, the project team determined to expand its engagement effort by integrating the SCI-Baltimore initiative into the many community initiatives underway in the city. In the spring of 2015, Baltimore experienced the ramifications of civil unrest in ways not experienced since 1968. Consequently, Baltimoreans entered into a period of soul-searching and reform-minded discussion – formally, informally, and via social media. With a non-incumbent mayoral election, various constituent organizations had been focusing in 2016 on the preparation of key priorities to ensure that new leadership be informed of and responsive to communities' needs. Several local convenings were underway or are being planned, and so rather than creating a wholly separate process for the SCI-Baltimore initiative, the project team determined it would be more effective and efficient to connect SCI-Baltimore to these ongoing discussions. This approach allowed SCI-Baltimore to promote a coordinated effort that built on community concerns and priorities voiced in real time. This effort became known as a "listening-to-the-listening" approach to community engagement. To put this idea into action, SDG partner organizations compiled a list of community initiatives that SCI-Baltimore could link to and work alongside. In many case, given its role as the local data intermediary, BNIA staff were already involved in these ongoing processes.

The project team connected with these community initiatives and attended associated events scheduled to take place in Baltimore. Working group members were also provided a running calendar of events so that they could participate and help raise awareness about the SDI-Baltimore initiative. Those who attended the events were tasked with documenting data on local development concerns, priorities, targets, and indicators that they heard and mapping those to the SDGs. Information from these community initiatives helped inform the project team's effort to take stock of existing city plans.

The "listening-to-the-listening" approach benefited the SCI-Baltimore effort by enabling the team to record and synthesize the diverse community concerns and desires being articulated by the public. The set of proposed indicators compiled for Baltimore also captured the sentiments of this synthesis.

2.3.4 Update to Baltimore's Sustainability Plan

Perhaps the most important and most similar concurrent effort was the update to the city's 2009 Sustainability Plan, which coincidentally

[3] "Which Goals Are You?" is an interactive quiz for users to better understand which of the 17 Global Goals they may be most passionate about. http://employers.globalgoals.org/

also began in early 2015. In large part as a response to the civil unrest that erupted in Baltimore in April of that year, the Baltimore Office of Sustainability (BOS) and Sustainability Commission were committed to ensuring local voices and particularly marginalized communities were included during the planning process for a more inclusive and equitable plan. In 2016, BOS launched the "Every Story Counts" campaign that gathered stories from residents who helped improve sustainability through their day-to-day actions in neighborhoods throughout Baltimore. The Sustainable Plan update also signed on and trained Sustainability Ambassadors to serve as outreach partners to collect more stories through the campaign as well as test and disseminate a survey that reached 1200 respondents.

In many respects, given the time and the subject matter, the SCI-Baltimore process and the Baltimore Sustainable Plan update should have been highly complementary if not fully integrated. Instead, the tension between local and global initiatives became acute rendering the need for more nuanced and deliberative approach by the SCI-Baltimore team to broker any kind of alignment. By way of example, below was one of the unforeseen initial reactions by BOS to the SCI-Baltimore process:

> The Sustainability Goals for our office are a bit different than the Sustainable Development Goals. We are incorporating elements of the STAR Community Rating System[4] into the structure of our plan and moving forward with extensive community outreach and input sessions over the next few months. I don't know if it makes sense to [incorporate] the SDG's … because it might confuse the process we're doing with the Baltimore City Plan update (Communication with the Baltimore Office of Sustainability, April 2016).

In response to this hesitation to combine efforts, BNIA and SDSN staff made back-of-the-envelope mapping between the STAR Community Rating System and the SDGs in 2016. This informal exercise and ongoing conversations between BNIA and the Baltimore Office of Sustainability spurred the Baltimore Community Foundation (BCF) to support efforts to identify alignment between the strategies in the new Sustainability Plan and the localization of the SDGs. After the SCI-Baltimore localization process had ended, in 2017, BCF provided a grant to BNIA to ensure clear linkages in terms of language, SDGs, and indicators to the city's Sustainability Plan which was ultimately adopted in January 2019.[5]

In the final version of the plan, the relevant SDGs are highlighted at the beginning of each chapter for each of the local goals (see Fig. 2.1). While this incorporation does send a clear signal to anyone reading the plan about the policy- and action-related connections to the SDGs, no further linkages were featured between the quantitative targets in the local plan to the global goals.

2.4 Stocktaking of Sustainable Development Plans, Initiatives, Goals, and Targets

One of the main objectives of the SCI-Baltimore process was to take stock of existing plans and initiatives in the city relating to sustainable development. With research assistance from SDSN and based on input from the local stakeholder convenings, a broad range of plans and documents were reviewed to assess whether their targets and goals were already aligned with the SDGs. Certainly, the city's 2009 Sustainability Plan figured prominently in the review, but several other city and statewide plans were also relevant such as the Baltimore Climate Action Plan, the Journey Home Plan (Homelessness), and Maryland Port Administration Environmental

[4] Many US cities like Baltimore have been part of the partnership that includes ICLEI – Local Governments for Sustainability, the US Green Building Council, the Center for American Progress, and the National League of Cities to address the needs of US cities, towns, and counties seeking a common framework for sustainability. The STAR Community Rating System was initially released in the fall of 2012 and not updated to respond to the UN SDGs until June 2016.

[5] "The 2019 Baltimore Sustainability Plan" https://www.baltimoresustainability.org/plans/sustainability-plan/

Fig. 2.1 Example of connecting local strategies to the global goals. (Source: The 2019 Baltimore Sustainability Plan, Chap. 5, "Human-Made Systems")

Strategy.[6] As mentioned previously, the stocktaking effort identified (a) partner organizations that could share tacit knowledge on development in Baltimore and collaborate on the SDG achievement effort and (b) indicators and measure data for SDG target tracking. The information yielded through the stocktaking effort grew with each consultative event as SDG partners convened and pooled information on their work and the work of others in the city, making the stocktaking exercise a "living" process to promote efficiency by building a coordinated SDG effort from existing sustainable development knowledge, resources, and activities.

Of course, given the comprehensive and interconnected nature of the SDGS, the breadth of stocktaking made clear that Baltimore was regulated and guided by many, many plans created by city and state governmental agencies in response to legislative and executive mandates. For example, the Maryland Department of Transportation annually prepares the Air Quality Attainment Report to ensure the regional compliance with the 1963 Federal Clean Air Act. Every 6 years, the city of Baltimore prepares a Comprehensive Economic Development Strategy (CEDS) to remain eligible for program funding from the Economic Development Agency (EDA). By 2016, the UN Sustainable Development Goals did not yet figure into existing regulatory mandates for cities nor did that come with substantial monetary resources to address the scale of the needs. Without these, for a distressed city like Baltimore, the SDG localization initiative hardly seemed worth the extra effort for executive branches of the government.

2.5 Developing the Preliminary Set of SDG Indicators for Baltimore

The second key objective of the SCI-Baltimore initiative was to identify locally relevant and useful indicators that could be used to set and track progress toward SDG-aligned targets. These indicators, developed out of the stocktaking exercise, considered stakeholder data needs for their own SDG-aligned development efforts and incorporated community aspirations. Community buy-in and collaboration with a wide range of stakeholders are essential to the success of the SDGs. By establishing indicators to measure progress toward development outcomes, accountability and transparency of public programs and non-profit initiatives is maintained, and results can be more effectively achieved. Support for expanded and improved data collection will bolster the city's existing efforts to eliminate poverty and homelessness, increase opportunities for employment and education, and protect the environment.

[6]See Appendix 1 (Existing Plans and Indicators) of the full report "Baltimore's Sustainable Future: Localizing the UN Sustainable Development Goals, Strategies and Indicators." https://www.ubalt.edu/about-ub/sustainable-cities/

2.5.1 Indicators for Baltimore—The *Vital Signs* Report

Defining and tracking indicators with community buy-in, of course, was not a new concept in Baltimore. Since 2000, the Baltimore Neighborhood Indicators Alliance (BNIA) at the University of Baltimore has served an alliance of groups and individuals in Baltimore dedicated to well-informed decision making for change. Since 2002, BNIA has published the annual *Vital Signs* report,[7] a compendium of over 100 community-based indicators for every Baltimore neighborhood. BNIA annually updates and provides the most current data as a part of *Vital Signs* and expands on existing data and indicators through a learning network of other cities engaged in the National Neighborhood Indicators Partnership (NNIP). Through ongoing and continuous consultation with neighborhood residents, leaders from across Baltimore, and data provider partners, BNIA designed its core functions based on Baltimore's need for a common way of understanding how its neighborhoods and overall quality of life are changing over time. The work illuminates changing conditions and provides a mechanism to hold Baltimore and all others who work, live, play, and invest in its neighborhoods accountable for positive growth.

For Baltimore, the SDGs provided a framework for evaluating the strength of the set of indicators included in *Vital Signs*. The exercise of mapping the *Vital Signs* indicators to the SDG targets presented gaps in both frameworks; the SCI-Baltimore initiative offered a means to address both local and global missing elements.

2.5.2 Preparing SDG Targets and Indicators of Success in Baltimore

Based on years of experience with local issues and local data, BNIA was in a good position to focus the SCI-Baltimore initiative on a proposed Baltimore SDG Index. To develop the global targets for the 17 SDGs and a series of progress indicators to track, indicators were evaluated according to these guiding principles:

- The data aligns with SDG targets *and* represents local priorities.
- Indicators reflect existing/parallel processes envisioning Baltimore's future.
- Data is accessible and actionable and from a valid, reliable source.
- Baseline measures are recurring in order to be tracked over time.
- Measures can help address disparities through disaggregation by race and by gender.

A total of 56 indicators across the 17 SDGs have been identified through a series of SDG Executive Team and working group meetings, where indicators were reviewed, added, and removed from the selection, and through the listening-to-the-listening effort. All of the indicators drew from open data sources including the aggregation of data in Baltimore's *Vital Signs* report. The proposed indicators outlined in the pages below can be used by decision makers and stakeholders in Baltimore to set quantitative values for local targets that align with the Global Targets, and they can be used to track progress toward achieving those targets leading up to 2030.

To gauge the relevance to the community and other stakeholders, the resulting set of 56 indicators were then presented to and scored by those representatives at the Opening Session for Baltimore Data Day 2016. For each proposed progress indicator, the following information was collated: baseline measure, baseline year, data source, participant score from Baltimore Data Day, and a graphic that visualizes 3–5 years of baseline data and a potential trend line based on the existing trajectory out to 2030. After the 56 indicators were selected, several local and regional organizations signed pledges to support the specific SDGs and indicators that reflected their work and values.[8]

[7] Baltimore Neighborhood Indicators Alliance-Jacob France Institute, *Vital Signs* Full Report, http://bniajfi.org/vital_signs/fullreport/

[8] See Appendix 3 (Baltimore SDG Letters of Endorsement) of the full report "Baltimore's Sustainable Future: Localizing the UN Sustainable Development Goals, Strategies and Indicators." https://www.ubalt.edu/about-ub/sustainable-cities/

2.5.3 Review of Baltimore SDG Indicators at Baltimore Data Day

Beginning in June 2016, the SCI-Baltimore project team solicited public feedback via an online survey of the final 56 indicators that had been identified through SDG-ET and working group meetings. The survey was promoted via social media as well as through a media press release issued by the University of Baltimore. Additionally, the team obtained feedback via the project website, which provides details on these indicators for public review. As another example of leveraging existing local processes, the results of the indicators compilation were prepared for Baltimore Data Day, an annual workshop hosted by the Baltimore Neighborhood Indicators Alliance to help local communities expand their capacity to use technology and data to advance their goals. At the seventh Annual Baltimore Data Day in 2016, community leaders, non-profit organizations, governmental entities, and civic-minded technologists came together to see the latest trends in community-based data, technology, and tools and learn how other groups are using data to support and advance constructive change.

An Opening Session on Sustainable Development, held on July 21, 2016, one day prior to the annual Baltimore Data Day workshop, provided an in-person opportunity for SCI-Baltimore partners and participants to provide additional feedback on the set of indicators. More than 130 people registered for the Sustainable Development Opening Session of Baltimore Data Day, which was hosted by the Federal Reserve Bank of Richmond's Baltimore Branch.[9] The general public was invited to provide comments on posters for each of the SDG indicators, using stickers to answer this question for each indicator: "Do you think a change in this indicator addresses the sustainable development goal?" (yes, no, maybe). The public engagement with the posters themselves provided visual ways for audience members to see how relevant the indicators were to participants. See adjacent photo example. The poster responses combined with the online survey results provided key feedback as to which indicators are deemed important to a broad spectrum of Baltimore stakeholders.

Responses were positive overall, with an average score among them of 4.2/5. Individually, the majority of proposed indicators scored either a 4 or a 5 on the scale provided. There were a few that scored 3 or less, and based on that feedback, the project team

[9]The Opening Session on Sustainable Development also featured keynote speeches by Professor Jeffrey Sachs and former Maryland Governor Parris Glendening.

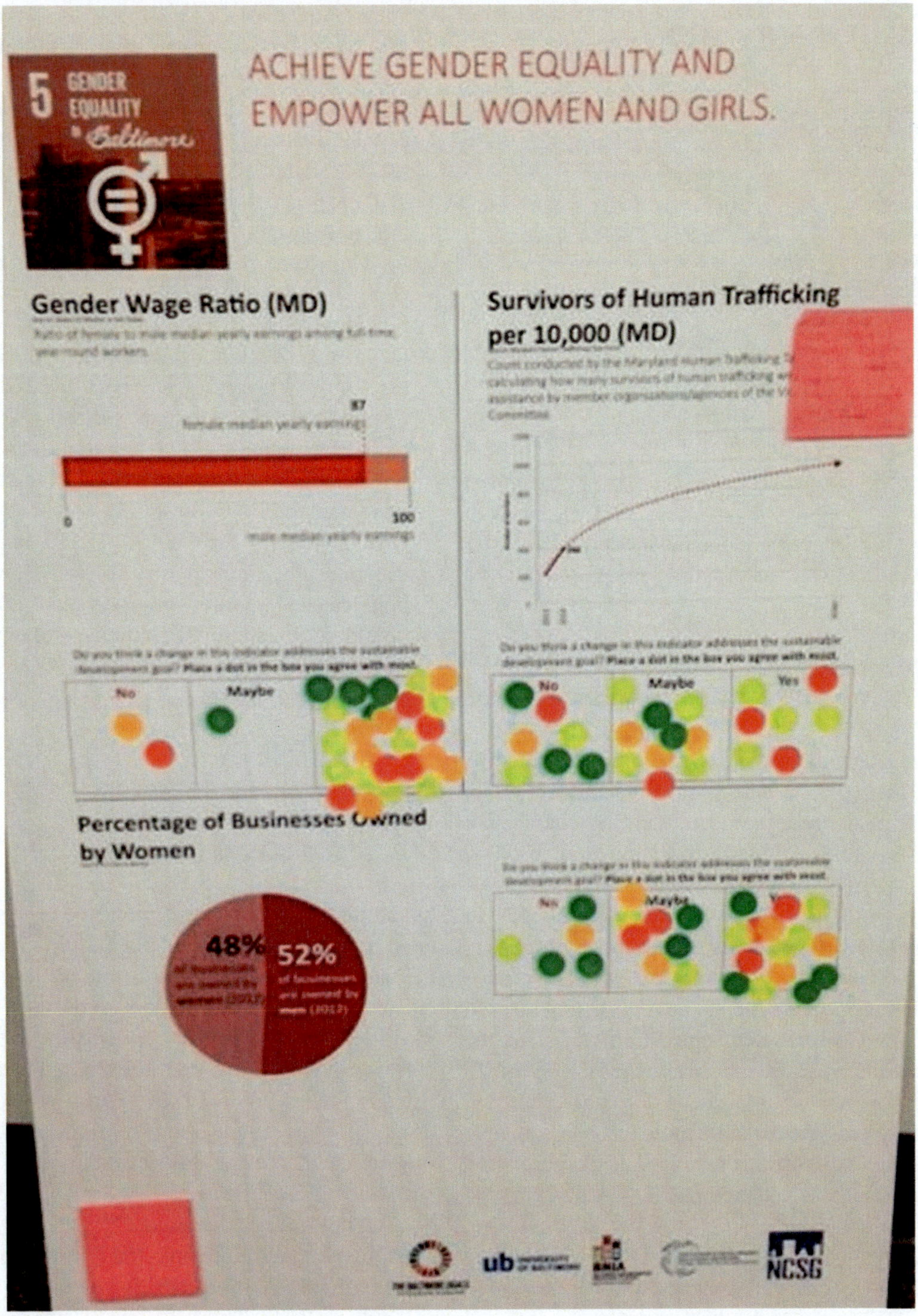

Fig. 2.2 Figure proposed SDG indicators for Baltimore in an interactive display at Baltimore Data Day (2016)

Table 2.1 Indicators receiving low relevance scores by attendees at Baltimore Data Day, 2016

Goal	Indicators receiving low feedback scores
5 Gender Equality	Survivors of Human Trafficking per 10,000 Residents
7 Affordable and Clean Energy	Total Electricity Consumption per Capita
9 Industry, Innovation and Infrastructure	Number of Utility Patent Grants
11 Sustainable Cities	Number of Days with Air Quality Index "Good"
13 Climate Action	Number of Excessive Heat Code Red Days
16 Peace and Justice	Percent Registered Voters Who Voted in the General Election

recommended that they should potentially be removed from the final list or revised (Table 2.1).[10]

2.5.4 New Local Indicators for SDG#16: Peace and Justice

Given the timing of the SCI-Baltimore initiative after the period of unrest in the city, review of the SDGs made clear that there were no quantitative measures to track progress toward a more just city. The Maryland Access to Justice Commission was reconstituted in 2015 as an independent entity to promote legal awareness, equal access to justice, and fair outcomes for all Marylanders who encounter the civil justice system. The commission focused on SDG Goal #16 and helped BNIA convene the Justice Indicators Roundtable for the SCI-Baltimore initiative to discuss methods for measuring and tracking progress toward a more just and equitable city. These discussions produced several proposed measures that would promote SDG #16 targets that are critical to achieving progress in Baltimore. However, some requisite data is not yet publicly available for calculating and monitoring these indicators. The proposed indicators are as follows:

- *State/Local Public Funding for Legal Aid for Eligible Clients*: Cost is often a prohibitive factor restricting a person's access to legal representation. This indicator is intended to capture availability of affordable legal counsel. Maryland Access to Justice Commission is in the process of procuring data to calculate this indicator.
- *Length of Time in Jail Pretrial for Misdemeanor Offenses:* Criminalization of poverty is a major problem. This indicator will track the prevalence of civil or misdemeanor cases that result in increased severity of legal consequences due to a defendant's inability to post bail or pay fines. BNIA continues to work with the State's Attorney's office to calculate this indicator.
- *Civil Legal Aid Attorney Ratio:* To calculate this ratio, the number of full-time-equivalent civil legal aid attorneys employed in Baltimore would be divided by the number of people in the state with incomes at or below 200% of the federal poverty level.

2.5.5 New Local Indicators for SDG#1 (Poverty) and SDG#3 (Health)

In addition to the indicators proposed to be calculated for Goal #16 (noted above), two other key indicators require dedicated resources to be calculated for Baltimore. These indicators were identified through consultations with the SDG-ET and working groups and the listening-to-the-listening effort. With funding from an international granting foundation, BNIA was able to prepare new indicators for Baltimore in 2017.[11]

- *Percent of Residents Earning a Living Wage:* A living wage is the hourly wage, a wage that is high enough to maintain a normal standard of living. In 2016, the living wage for a single adult in Baltimore is $12.33. Using a living wage methodology established by the

[10] "Data Day Scores" were calculated as follows. Feedback for each indicator was weighted accordingly: yes = 5, maybe = 3, and no = 1. Scores were summed for each "dot" or "vote," and that sum was divided by the total number of votes.

[11] Seema D. Iyer (2017) Localizing the UN Sustainable Development Goals in Baltimore: Next Steps Towards Implementation https://www.unsdsn.org/news/2017/12/20/localizing-sdgs-in-baltimore-next-steps

Massachusetts Institute of Technology (MIT), the results show that households with two adults were far more likely to earn more than the living wage than households with only one adult. For one-adult households with children, the impacts are even more severe. Only 13% of one-adult/one-child households earn more than the living wage; only 6.4% of one-adult/two-children households earned more than the living wage. The SCI-Baltimore initiative identified this indicator as relevant to setting and tracking SDG#1 targets for Baltimore.

- *Life Expectancy by Race:* Life expectancy (the average number of years a newborn can expect to live), assuming he or she experiences the currently prevailing rates of death through their lifespan, would be the premier indicator for tracking the health of Baltimore residents. BNIA worked with the Baltimore City Health Department to calculate life expectancy by neighborhood and disaggregated this data by race. In 2017, white Baltimoreans (76.1 years) lived an average 6 years longer than black Baltimoreans (70.9 years). This indicator is relevant to setting and tracking SDG#3 targets for Baltimore.

2.6 Efforts to Promote Lessons Learned from Baltimore

The lessons from the SCI-Baltimore SDG localization effort helped put Baltimore on a worldwide platform, in large part through the connections and promotion by the Sustainable Development Solutions Network (SDSN). Beginning as early as 2016, representatives from the localization team were invited to attend meetings at the United Nations, the State Department, the Brookings Institution, and national organizations interested in supporting the SDGs such as the Council on Foundations. The city's Office of Sustainability was invited to participate in Habitat III in Quito in 2016, although was unable to attend. The new Baltimore mayor, Catherine Pugh, participated on a 2018 panel called "Localizing the SDGs: Achieving the Global Goals Through U.S. Cities" at the winter meeting of the US Conference of Mayors, with Mayor Buddy Dyer of Orlando and Mayor Mitch Landrieu of New Orleans. Additionally, several news outlets featured the efforts in Baltimore to global audiences.[12]

The work on development of localized indicators from Baltimore also figured prominently in the development of the US Cities Index (Espey et al. 2018). However, while Baltimore may have been at the forefront immediately after the SDGs were adopted in 2015, the city can hardly be seen as a beacon just 3 years later. With a new federal administration deprioritizing federal efforts to track the SDGs nationally and a lack of continued resources to support localization, only tepid connections to the global goals can been seen in Baltimore today.

2.7 Conclusions

For Baltimore to be one of the first cities to participate in USA-SCI represented a moment of opportunity to be at the forefront of a global conversation. Given the historical moment, however, the project team had to thread the SCI-Baltimore initiative into local issues, process, and realities to raise awareness about the global goals and to make translations about their relevance locally. The work of the initiative, therefore, was much more about listening, educating, and "mapping" to more familiar frameworks than about the technical needs for calculating the proposed set of local indicators. Some positive and long-lasting benefits have accrued to Baltimore. The global goals have been incorporated into the city's 2019 Sustainability Plan, and several new indicators particularly regarding equity and justice have been calculated for Baltimore. There are also examples of Baltimore stakeholders participating in other programs[13] created to respond to the

[12] See, for example, "How Baltimore Is Using the Sustainable Development Goals to Make a More Just City" by Carey L. Biron, Citiscope (March 2017).

[13] Baltimore is one of only 8 US cities in the European Union's International Urban Cooperation (IUC) program activities to foster city-to-city knowledge-exchange for sustainable development (SDG #17—Partnerships).

SDGs. More comprehensive and more target-based connections to the global goals, however, seem unlikely today.

2.7.1 Generalizations

From the Baltimore experience, three key issues arose that could both hinder and help other North American cities from focusing on the SDGs.

1. *Lack of resources and leadership at all levels of government to support adoption of the SDGs*. The vacuum of local leadership precisely during the moment of localization in Baltimore is of course a rather obvious impediment in the localization effort. What was less overt but equally important was the lack of Federal and State leadership among US and Maryland agencies engaged in urban policy. The US State Department and particularly the Office of the Chief Statistician under the Obama administration certainly helped keep the data collection and voluntary reporting at the forefront through the Data Revolution for Sustainable Development Initiative. However, many of the key agencies that cities interact with more routinely, such as HUD, EPA, or DOJ, were not disseminating similar messages or guidance about how to align the SDGs to local priorities either from a regulatory perspective or via resource allocation and funding.
2. *Lack of support for or awareness of the SDGs from urban entities with similar missions*. With the UN adoption of the SDGs in 2015, what may have been a seminal moment globally had hardly made an impression on local organizations or professionals involved in US-based sustainability movements. The fact that Baltimore's own Office of Sustainability (BOS) was leery of adopting a "non-local" approach to sustainable development was certainly an unforeseen barrier to localization. Staff from BOS were well-connected to networks such as the STAR Community Ratings and the Urban Sustainability Directors Network (USDN), but at the time, guidance on the SDGs had not yet permeated within these spheres.
3. *Local indicator projects are in the best position to help translate the relevance of the SDGs to urban communities*. BNIA was not the first partner chosen by SDSN to help spearhead the SCI-Baltimore initiative; however, having the local community indicators project be a part of the SDG localization process became critical to ensuring Baltimore saw the effort to fruition. Given its long-standing mission to improve quality of life in distressed communities, BNIA already had deep connections with local governmental agencies as well as neighborhoods impacted by the unrest which helped ensure inclusive working group participation and an effective listening-to-the-listening approach. BNIA was also able to quickly align existing, locally relevant indicators to the SDG targets and identify gaps in data that allowed stakeholders to focus on ways to measure equity and justice.

2.7.2 Recommendations

Having gone through an intensive process to localize the SDGs in Baltimore, any attempts in other cities would benefit from these internal and external supports to help convey the potential benefits for aligning with the global goals:

1. *Message needs to come from the top*. The ambitious and comprehensive nature of the global goals will need to be addressed by all levels of government, with most of the responsibility resting with the executive offices and agencies (Kingsley 2017). Clear and reinforced language within existing regulations and funding resources could help agencies better understand that tracking the SDGs fits within ongoing workloads and reporting practices. This would require incorporating language within agency-promulgated rules and regulations and in rare cases within legislation. This important task requires leadership from within

government and potentially advocacy from constituencies.

2. *Existing networks promoting urban sustainable development need to be involved.* The SDGs represent a new framework for thinking about and quantifying sustainable development; however, they do not represent new issues for US cities. For many years, grassroots efforts have grown in the USA to address urban sustainability, so making clear connections between existing priorities within organizations involved in any/all aspects of sustainability will ensure buy-in from stakeholders already at the forefront of sustainable development in North America.
3. *Local data is critical for tracking the SDGs.* Having an existing repository of local data collected by community indicator projects like BNIA enabled speedier collection of baseline data relevant to the global goals. In fact, based on a recent report by the Urban Institute, whereas 66% of the SDG targets could be measured using national datasets alone, 81% of the targets were measurable if supplemented with local data (Greene and Meixell 2017). Of course, local indicators projects provide far more to urban communities than just the data alone; they offer training and education to multi-sector stakeholders as well as continuous integration of new local datasets that arise from local policies and administration. They are nimble enough to help map local realities to global issues using the common approach of the quantitative targets, which is precisely what the SDG localization process in Baltimore helped reveal.

References

Biron CL (2017, March) How Baltimore is using the sustainable development goals to make a more just city. Citiscope

Espey J, Dahmm H, Manderino L (2018) Leaving no U.S. city behind: the U.S. cities sustainable development goals index. Sustainable Development Solutions Network. https://www.sdgindex.org/reports/2018-u.s.-cities-sdgs-index/

Greene S, Meixell B (2017) Hacking the sustainable development goals: can US cities measure up? Urban Institute. http://www.urban.org/research/publication/hacking-sustainable-development-goals

Iyer SD, Davis A (2017) Localizing the UN sustainable development goals in Baltimore: next steps towards implementation. Sustainable Development Solutions Network. https://www.unsdsn.org/news/2017/12/20/localizing-sdgs-in-baltimore-next-steps

Iyer SD, Howland A, Ruckstuhl S, Sheehan IS, Samuelson L, Bowman L (2016) Baltimore's sustainable future: localizing the UN sustainable development goals, strategies and indicators. University of Baltimore. https://www.ubalt.edu/about-ub/sustainable-cities/

Kingsley GT (2017) A broader view of the data revolution and development agenda. Urban Institute. https://www.urban.org/research/publication/broader-view-data-revolution-and-development-agenda

Seema D. Iyer, Ph.D. (co-editor), oversees the Baltimore Neighborhood Indicators Alliance-Jacob France Institute (BNIA-JFI) at the University of Baltimore (UB), which has been dedicated to providing reliable, community-based indicators for Baltimore's neighborhoods since 2000. BNIA-JFI is the local partner of the Urban Institute's National Neighborhood Indicators Partnership, a network of cities that provide longitudinal data on demographics, housing, crime, education, and sustainability. Under her leadership, BNIA-JFI has received several awards for its role in increasing accessibility to indicators for organizations such as the Community Indicators Consortium (2016), Harvard Ash Center (2017), and the Association of Public Data Users (2018). Since 2015, she has been involved in the city of Baltimore's localization of the SDGs and has advised cities around the world on how they can work with indicator programs to help monitor progress. She is also director of UB's undergraduate program in real estate and economic development. She holds a Ph.D. in urban and regional planning from the University of Michigan.

3 Los Angeles: Measuring Its Ambition to Achieve the SDGs

Tony Pipa

3.1 Introduction

On February 15, 2018, in a speech at Occidental College, Mayor Eric Garcetti announced in a public speech that the city of Los Angeles would pursue achievement of the Sustainable Development Goals (SDGs).

Mayor Garcetti's pronouncement was unusual for several reasons. The SDGs, a set of 17 ambitious goals to end poverty and promote equity, strengthen peace and security, and enhance environmental sustainability, were agreed by countries at the United Nations as part of the 2030 Agenda for Sustainable Development.[1] While the agreement is voluntary, with no legal force of compliance, the expectation of leadership and accountability clearly rests with the federal government. Los Angeles city government has no specially designated role and was not party to the agreement.

One of the goals, SDG 11, focuses on the importance of cities to sustainable development, committing to "make cities and human settlements inclusive, safe, resilient and sustainable." The targets from this goal include adequate and affordable housing, accessible transportation, participatory planning, improved resilience against catastrophe, protection of cultural heritage, and reduction in environmental impact of cities. The main intent of this goal is to focus the attention of national governments on the implications of urbanization and the importance of managing it well in order to achieve sustainable development. It also provides the basis for a discourse between the different levels of government in managing that growth sustainably.

Mayor Garcetti's vision for Los Angeles, however, extends beyond the dimensions outlined in SDG 11. It commits Los Angeles to implementing the SDG agenda in total and places the city in the central role of protagonist – prioritizing, managing, and measuring the social, economic, and environmental progress its leadership can deliver for its neighborhoods and citizens.

The targets and metrics of the SDGs agreed upon at the UN are set at the national level. Applying them to Los Angeles thus requires adaptation and judgment. The time frame – achievement of the goals is to occur by 2030 – extends beyond the electoral cycle. The Los Angeles commitment could extend through the tenure of three more mayors.[2]

Mayor Garcetti's commitment placed Los Angeles in rare company nationally. Such a pub-

[1] *Transforming Our World: The 2030 Agenda for Sustainable Development*, A/Res/70/1, UNGAOR, 70th Session (2015)

[2] Mayor Garcetti is term-limited to two terms and started serving his second 4-year term in 2017.

T. Pipa (✉)
Brookings Institution, Washington, DC, USA
e-mail: TPipa@brookings.edu

D. B. Abraham, S. D. Iyer (eds.), *Promoting the Sustainable Development Goals in North American Cities*, Sustainable Development Goals Series, https://doi.org/10.1007/978-3-030-59173-1_3

lic and specific mayoral commitment to the SDGs is uncommon in the USA and immediately put Los Angeles in the forefront of local implementation globally. Mayor Garcetti committed the city to the SDGs in a public speech to his constituents, rather than a global forum of international leaders, the more usual platform for local leaders to endorse the SDGs.

Adopting the SDGs is easy. The aspirations that illuminate the agenda are attractive to any locality and make for soaring political rhetoric: an end to poverty and its related indignities; an increase in quality jobs, needed infrastructure, and affordable housing; and a reduction in inequality, all while ensuring a sustainable environment for future generations.

Serious implementation to achieve the SDGs is much more difficult. The 17 goals cover the full breadth of development, with 169 targets, most of them time-bound and outcome-oriented. The agenda exposes interdependencies among different dimensions of development and requires progress on multiple fronts simultaneously. Charting progress toward its goals with publicly visible data exposes the full breadth of a government's successes and shortfalls. The SDGs also set expectations at a level that goes beyond the capability of a local government's resources, requiring city officials to attract and align the contributions of multiple major stakeholders.

The commitment to pursue the SDGs reflects an ambitious political vision for Los Angeles. Since the SDGs were designed for national implementation and measurement, cities interested in the SDGs as the basis for local progress must blaze their own trail as they align community plans against the goals. This case study reviews the first year of LA's implementation of the SDGs. It explores the incentives for pursuing the goals and analyzes the steps that Los Angeles has taken to align with the SDGs and measure its social, economic, and environmental progress. It surfaces lessons for other cities considering the use of the SDGs as a blueprint for community progress and ends with recommendations for LA's next steps.

3.2 Background: City of Los Angeles

Los Angeles is the second largest city in the USA, with a population of more than four million people within its city limits. It is one of the most diverse metropolitan areas in the USA. Its city government manages a budget of $9.9 billion[3] and employs approximately 64,000 people. In 2017 it was estimated to have the third largest metropolitan economy in the world, roughly equivalent to the size of Turkey's economic output.[4]

While a growing, thriving city, Los Angeles also faces tough urban challenges. Homelessness in the city and county increased by 75% over the 6 years leading into 2018, with the city ranking near the bottom in sheltering its homeless relative to its US peers.[5] The city's poverty rate in 2017 topped 20%,[6] more than eight percentage points above the country's poverty rate. In 2018, it reported a record 87 consecutive days straight of unhealthy ozone levels,[7] and it has a history of ranking first globally for traffic congestion in an annual ranking of gridlocked cities.[8]

The city government does not have primary managerial control for public services with regard to major parts of the SDG agenda. Los Angeles County, within which the city sits, adopted a budget of $32 billion for 2018–2019 and operates the public hospitals and clinics. The

[3] City of Los Angeles, Open Budget. 2019. Retrieved from http://openbudget.lacity.org/#!/year/default

[4] How the Economic Power of American Cities Compares to Countries (2017, November 12). Retrieved from https://howmuch.net/articles/the-economic-size-of-metro-areas-compared-to-countries

[5] Holland, G. (2018, February 1). L.A.'s homelessness surged 75% in 6 years. Here's why the crisis has been decades in the making. *The Los Angeles Times*

[6] US Census Bureau, Population Estimates 2018. Retrieved from https://www.census.gov/quickfacts/losangelescitycalifornia

[7] Barboza, T. (2018, September 21). 87 Days of Smog: Southern California Just Saw Its Longest Streak of Bad Air in Decades. *Los Angeles Times*

[8] Los Angeles Tops INRIX Global Congestion Ranking (2018, February 5). Retrieved from http://inrix.com/press-releases/scorecard-2017/

Los Angeles County Department of Health Services is the second largest municipal health system in the country. The Los Angeles Unified School District oversees public education in the city but also serves 26 other cities and unincorporated areas of Los Angeles County and even employs its own police force separate from the Los Angeles Police Department. Multiple levels of governance create challenges for driving shared progress on key social determinants.

3.3 Committing to the SDGs

The commitment by the city of Los Angeles to pursue the SDGs was a convergence of several factors, including opportunistic philanthropic leadership, a globally minded mayor who prizes evidence-based policy, and the city's designation as host of the 2028 Summer Olympic Games. The US federal government, as of 2019, has issued no plan to implement the SDGs and is providing no leadership, encouragement, nor resources to any stakeholders for SDG implementation. The effort by Los Angeles to commit to and adapt the SDGs for their own local purposes has been bottom-up, generated and supported by local stakeholders, offering a value proposition aligned with local political and programmatic priorities.

3.3.1 Key Factors

Philanthropic Leadership: The idea surfaced with the Conrad N. Hilton Foundation, a family foundation with interests both local (e.g., ending homelessness in Los Angeles County) and global (e.g., the worldwide elimination of trachoma). Ed Cain, then Vice President for Programs, had previously served as a country Resident Coordinator within the UN system. He realized that, unlike the predecessor Millennium Development Goals (MDGs), the SDGs are to be universally applied by both high-income and low-income countries, no matter the level of development.

In conjunction with other local philanthropies, the Hilton Foundation had supported *A Portrait of Los Angeles County*,[9] a report based on quantitative community indicators that was launched in November 2017 by Measure of America. Measure of America used a methodology based on the UN's Human Development Index to explore how the county's residents were "faring in terms of well-being and equity," publishing a ranked index for the 106 cities and unincorporated areas of Los Angeles County. The portrait contained a "Global Goals Dashboard," a distilled version of the SDGs with associated indicators specific to Los Angeles County. For the foundation, the experience highlighted that collaboration with its counterparts on tough social issues could be enhanced if they agreed upon and used a common framework.

Impressed with the relevance of the findings and interested in exploring the potential for the SDGs to help the city drive social and environmental priorities, the foundation's leadership approached the Mayor's Fund for Los Angeles, an independent and non-partisan organization that pools and leverages private financial resources to help the city take on challenging priorities. Together, the Mayor's Fund and the foundation found receptivity from key city officials, such as the city's Chief Sustainability Officer and its Deputy CIO, who helped elevate the idea to the mayor.

Mayoral Leadership: They were engaging a mayor with a strong international outlook, intent on establishing Los Angeles as a leader on the global stage. Just a few months before the launch of *A Portrait of Los Angeles County*, Mayor Garcetti had named Ambassador Nina Hachigian as the city's first-ever Deputy Mayor for International Affairs, to oversee a newly created Office of International Affairs. The mayor's international perspective, grounded in his experience as a Rhodes Scholar and former professor of diplomacy and world affairs, facilitated a willingness to connect the globally agreed SDGs with his local political agenda to make life better for his city's residents.

[9] Measure of America. *A Portrait of Los Angeles County.* (2017, November). Retrieved from https://ssrc-static.s3.amazonaws.com/moa/PoLA%20Full%20Report.pdf

A Unifying Event: Around the same time the International Olympic Committee awarded the 2028 Summer Olympic Games to Los Angeles. The expiration of the SDGs in 2030 aligns nicely with the timing of the 2028 Los Angeles Olympics, allowing the city to promote its efforts to ready the city for the Olympics as a simultaneous exercise in advancing sustainable development.

Data-driven Decisionmaking: The mayor also possesses a strong orientation for using data and goals to drive progress. He had made an SDG-like commitment in 2017 to decrease the number of unsheltered Angelenos by 50% in 5 years and functionally end homelessness in 10 years. This pledge demonstrated the mobilizing effect of aligning policy and budget against a publicly accountable goal. The mayor's political focus has helped produce a county bond for $355 million annually for services and programs, a city bond for $1.2 billion for supportive housing, and an executive directive to expedite the process of standing up temporary shelters.

Implicit Alignment: Indeed, the city's policies and plans already mirrored many of the priorities reflected in the SDGs. In April 2015, the mayor had released *Sustainability City pLAn*, which sought to integrate and measure environmental health, equity, and economic near-term and long-term outcomes. The city was also in the closing stages of finalizing *Resilient Los Angeles*, a plan developed in conjunction with Rockefeller Foundation's 100 Resilient Cities initiative. The city had made a commitment in 2017 to update all of its 35 community plans within 6 years through neighborhood-level consultations, and the Mayor's Dashboard was already providing real-time data measuring the city's performance across a number of issues and sectors.

University Support: The pieces fell into place with the addition of university partners. Occidental College offered financial resources, faculty involvement, and assistance from students. Ultimately faculty and students from ASU, UCLA, and USC would also participate. These university teams provided capacity for mapping existing city policies and metrics against the SDGs, undertaking labor-intensive analysis that might otherwise have taxed city staff. With a supportive mayor, the Hilton Foundation provided funds to the Mayor's Fund for a staff person in the mayor's office to take on the responsibility for coordinating the city's efforts on the SDGs.

3.4 Aligning to the SDGs

What does "implementation" of the SDGs entail? For the countries that agreed to the 2030 Agenda, the UN resolution emphasizes the importance of "cohesive nationally owned sustainable development strategies, supported by national financing frameworks," as well as regular review of progress using "a set of global indicators."[10]

The goals, targets, and indicators are set at the national and global levels. Using the data as a basis, countries develop strategy and financing frameworks according to their national circumstances. They then annually report national data to the UN Statistical Commission based on a standard set of indicators. Countries also voluntarily make Voluntary National Reviews (VNRs) at the United Nations, offering a self-assessment of national progress and presenting their plans to reach the SDGs.

However, officially determined and universally accepted SDG targets for local purposes do not exist. There is also no formal set of indicators or official forum for reporting local SDG progress.

Cities are thus faced with the prospect of creating their own proxies for the national targets and indicators, especially as it relates to their own specific context and the data that they have available. There is no straightforward "trickle-down" from the national to the local – cities must make decisions at every juncture. For example, Los Angeles could set target 1.2, a 50% reduction in poverty, at the nationally mandated poverty line or a poverty line that is more in line with the cost of living in the city.

[10] *Transforming Our World: The 2030 Agenda for Sustainable Development*, A/Res/70/1, UNGAOR, 70th Session (2015), pp. 28

In addition, cities face jurisdictional issues. Depending on local and national governance structures, they may not have primary public administration responsibilities for certain parts of the SDG agenda.

Cities are thus faced with a more complicated task. They are judging the extent to which targets are relevant to local circumstances while assessing the city's capabilities to achieve them. They must also choose the indicators that will measure their progress. These tasks are in addition to strategy, budgets and financing, and ongoing reporting.

From a pragmatic perspective, in the absence of a national mandate or the prospect of federal funding, the incentives and pay-off must outweigh the investment of the city's time and resources. Otherwise it is likely to lose interest in using the SDGs as a blueprint.

With these considerations, implementation of the SDGs for cities can generally be viewed along five lines of effort: (1) awareness, (2) alignment, (3) analysis, (4) action, and (5) accountability.

General awareness about the SDGs in the USA remains fairly low and generally benefits from a champion (in this case the Hilton Foundation) or a campaign to enable greater attention to the advantages that the SDGs might provide. After awareness captures the attention of key leadership, a process of alignment situates the city's priorities, strategies, policies, directives, initiatives, and activities within the aspirations and intent of the SDGs. It also identifies the indicators that the city will use to measure its progress toward the targets it has determined to be relevant to its circumstances. Subsequent analysis enables the city to build from that foundation, identifying where policy gaps or opportunities exist and have important implications for its ability to successfully reach the targets. Ideally that will lead to prioritizing and taking new action, through policy or budget proposals, public-private initiatives, new types of financing, and citizen and stakeholder engagement. Publishing reports or real-time data that measure progress provide accountability. These five lines of effort are often iterative, constituting a mutually reinforcing cycle as a city refines and deepens its activities.

The process of implementation entails (1) decisions about the relevance of the SDG targets to a local context, (2) the extent to which the city will customize the framework for its own purposes, and (3) the indicators and data sources it will use to assess progress. This is where Los Angeles focused much of its effort in 2018.

The process includes choices about policy ambition and choices about metrics that indicate progress. While these are related, they are not the same.

3.4.1 Policy Alignment

Policy alignment takes place on a continuum. For some cities, an existing strategy may act as the cornerstone, which the city maps to appropriate targets and priorities within the SDGs. The SDGs are thus viewed through the lens of current city priorities.

Alternatively, a city can start from the perspective of the SDGs and conceivably create a development strategy with targets and goals taken directly from the framework. In such a scenario, the SDGs act as the template for the city strategy.

The approach by Los Angeles falls somewhere in the middle of the continuum, mapping existing plans and policies to *and* from the SDGs.

This took place against the backdrop of translating the agenda to the city level, with Los Angeles judging the relevance of a specific target to its local context, and deciding whether it makes sense to make an adjustment.

Los Angeles city staff made it a core principle to be holistic and comprehensive in its approach. Each of the 169 SDG targets was tested against the city's plans and activities. The university student teams did this through a desk review and analysis that included the city's sustainability plan, its resilience strategy, department plans and activities, and the city's budget.

As a first step, they tested the applicability to the city of each target as written and agreed in the United Nations resolution. This identified a subset of 69 SDG targets that needed no change in language or quantitative ambition to be applicable to Los Angeles.[11]

For the remaining 100 targets, the teams adhered to several principles to guide the analytical process: (1) make the fewest changes possible to render the target applicable, (2) remain as faithful as possible to original intent, and (3) reflect the city's values, realities, and defined ambitions, demonstrating a strong commitment to inclusiveness and leaving no one behind.[12]

Victory was not automatically declared on any target. For example, SDG target 1.1 focuses on ending extreme poverty. Using the global standard of $1.90/day as outlined in the SDGs, the city has a strong case for claiming it has achieved the target. However, given the emphasis on grounding the exercise in the city's reality, the team recommended substituting an income level ($33/day) that would be reflective of extreme poverty in Los Angeles.

This means ambition sometimes exceeds the SDGs. For example, the ratios for the recommended Los Angeles targets on maternal mortality and preventable child deaths under five are far lower than the SDG targets: for maternal mortality (target 3.1), Los Angeles is aspiring to 5 per 100,000 live births, versus the SDG target of 70 per 100,000 births; for under five mortality (target 3.2), Los Angeles is aspiring to 4 per 100,000 live births, versus the SDG target of 25 per 100,000 live births.

Setting ambition at these higher levels facilitated the creation of a Los Angeles-specific Leave No One Behind agenda. African-American women, for example, experience much higher rates of maternal mortality in Los Angeles than Caucasians. To reach the recommended ratio, Los Angeles will need to disaggregate data among demographic groups and develop specific strategies to meet the needs of African-American women.

[11]Some of these targets incorporate international agreements or conventions to which Los Angeles is not a party, but by which the city can still abide.

[12]"Leave No One Behind" is often used as a shorthand for the imperative implicit in the SDGs that countries and stakeholders must reach their most vulnerable populations in order to achieve many of its targets.

Ultimately 156 targets comprise the recommended Los Angeles SDG framework. Thirteen targets were set aside, most of them means of implementation targets focused on resource or knowledge exchange between developed and developing countries, where legal structures or original intent was not meaningful to Los Angeles. The team recommended adding one target not a part of the SDGs, a target focusing on equity for LGBTQI, to extend the equity dimensions of the SDGs to populations important to Los Angeles (in similar fashion it made slight language modifications to expand the inclusivity of some targets).

As of February 2019, policy owners within Los Angeles city government are validating each recommendation relevant to their areas of responsibility. Their agreement on the policy recommendations will be critical for credibility, to ensure that the proposed framework is an accurate representation of Los Angeles ambition and context (e.g., does the proposed measure of $33/day accurately reflect a level of extreme poverty for Los Angeles?). Those policy owners will also provide important guidance and expertise on choosing appropriate indicators to measure the city's progress and performance.

3.4.2 Indicator Alignment

As a globally agreed and vetted agenda, the SDGs offer the promise of comparability in measurement. Their predecessors, the Millennium Development Goals (MDGs), were hailed for their positive impact in helping a diverse set of stakeholders, including countries, aid agencies, philanthropies, and implementing partners, agree upon and use a standard set of metrics related to the specific targets.

The breadth and depth of the SDGs pose significant challenges in this regard. The official global indicators used at the national level, developed by the UN Statistical Commission (UNSC) after significant consultation, are classified into three tiers. These denote their level of readiness and availability.

Tier I indicators reflect an internationally accepted methodology, with data produced by 50% of countries. The indicators in Tiers II and III have weaknesses either in methodology or availability, or both. As of December 31, 2018, two full years into implementation of the SDGs, just 45% of the official indicators for use at the national level are classified as Tier I.[13] This is after the international community has been working on the MDGs for 15 years, spent 3 years developing the SDGs, and has been engaged in implementation for 2 years.

Adapting SDG measurement to the local level increases the degree of difficulty. Data quality challenges similar to those experienced with the UNSC indicators not only exist locally, but for most municipalities, they will be more pronounced. These challenges are complicated by the reality that no officially determined SDG metrics exist for local purposes and that localization of the agenda can take many forms. A key issue relates to the balance between standardization and customization: To what extent is it important that the common language of the SDGs translates into common measurement across different cities?

The points of reconciliation among standardization and customization depend in part upon audience and objective. A city may focus its efforts one way if it places high value on the ability to compare progress against counterpart cities across the world. It may take other approaches if it sees the SDGs as a common denominator among different levels of government at the county, state, and national levels, or if it is primarily interested in using the SDGs as a common platform to mobilize action among community stakeholders.

As the city of Los Angeles selects indicators to measure progress toward the proposed 156 targets of its localized framework, there are multiple options from which to draw. Its Mayor's Dashboard provides regularly updated data on close to 200 indicators, measures, metadata, and charts. In 2018, the city entered into an agreement with the World Council on City Data (WCCD) to become one of eight local data hubs for sharing information based on WCCD's open data standards. There may also be the opportunity to localize selected national SDG indicators, especially if local data can be disaggregated from the same sources the US government is using to report national metrics to the UN through its online reporting portal.

The WCCD partnership offers promise in providing a common basis for comparison to other cities. In 2014, after years of consultation with cities worldwide, the organization was instrumental in publishing ISO 37120, a standard set of 46 core and 54 supporting indicators and related methodologies to measure the sustainable development of communities. Certification against the ISO standard requires third-party verification, ensuring a rigorous application of the methodology and a high quality of reporting. This enables a high degree of comparability for cities reporting against the standard.

Though the ISO standard was developed before the SDGs, WCCD recognized that the indicators cover similar social, economic, and environmental dimensions. WCCD now publishes an annual report of reporting cities with the ISO indicators mapped against the SDGs.

The alignment draws an association between each indicator in ISO 37120 to any SDG target where that indicator might provide relevant insights. Thus indicators may be used more than once, and more than one indicator may be associated with an SDG target.

While this approach helps draw an aggregate picture of progress, its usefulness in helping a city measure performance against particular SDG targets, especially for managerial purposes, seems limited. This is not surprising, as the ISO certification was not designed specifically for this purpose.

The set of 100 ISO indicators also leave gaps in coverage over the entire breadth of SDG targets. A rough analysis suggests that ISO indicators can provide a direct measure that corresponds to the specificity of a target, at a level equivalent

[13] IAEG-SDGs. *Tier Classification for Global SDG Indicators*. Retrieved from https://unstats.un.org/sdgs/iaeg-sdgs/tier-classification/

to the UNSC indicator at the national level, for about 15% of the agenda.

Los Angeles was one of the first cities worldwide to achieve platinum certification by WCCD for ISO 37120. While its continued reporting will be critical in measuring its progress against comparable cities in the world and identifying counterparts with whom to share best practices and innovations, the indicators constitute only a subset of the wide range Los Angeles will need in order to measure its progress on the targets in its proposed framework.

Given the intense localization of its SDG approach, Los Angeles will need to craft or identify unique indicators as it seeks to measure its performance with maximum rigor. Several key principles could help improve comparability as choices on data metrics and methodology are made: (1) apply WCCD indicators where directly corresponding, (2) identify relevant indicators that utilize data from nationally available sources (e.g., census bureau data), and (3) identify indicators with internationally or nationally accepted methodology and data.

3.5 Generalizing the Experience of Los Angeles

Taken together, the factors underpinning the city's commitment may seem unique to Los Angeles, an unusual mix of political, substantive, and personal interests. Yet the considerations mirror those of any city weighing the value proposition of the SDGs against the investment of time and resources in adapting them. Five key principles emerge based on the Los Angeles experience:

- *The SDGs represent a new language more than new priorities.* In many respects, Los Angeles found it was already working on the SDGs. The plans and priorities in place or under development contained most of the issues and aspirations reflected in the 2030 Agenda. Most communities are not starting from scratch. The SDGs offer a framework to help Los Angeles and other cities integrate their diverse activities and priorities and incorporate them into a comprehensive vision.
- *They offer a global outlook to local priorities.* Being an Olympic city is just one manifestation of the global purview of Los Angeles. Situating its local experiences and aspirations within an internationally recognized and agreed framework reinforces this outlook. Cities seeking to demonstrate a global perspective are likely to be similarly attracted, given the chance it offers to articulate how local progress demonstrates a measure of responsibility for global solutions. In today's interconnected world, the SDGs also give Los Angeles a common language to share aspirations, challenges, practices, and performance with counterpart cities across the world.
- *Political and technical comfort with goals and data is necessary.* Los Angeles' commitment to the SDGs builds upon existing data collection and reporting systems and even publicly announced benchmarks. Its pursuit of the SDGs puts the city in the forefront of counterparts nationwide creating evidence-based policy. At the same time, the SDGs invite public accountability and transparency, so elected and senior officials must demonstrate a willingness to expose their credibility and reputation based on their progress in reaching the targets.
- *Partnerships enable the agenda.* In many respects, local implementation of the SDGs entails complexities that are not present at the national level that at minimum require an investment of staff and time. The commitment by Los Angeles benefited from an injection of capacity, resources, and leadership offered by the Hilton Foundation, as well as the various university partners. These were instrumental in helping Los Angeles get underway. While the city has committed to covering the staff expense past the Hilton Foundation's 2-year commitment, further expanding the participation of external stakeholders will play an important role in accelerating and advancing progress toward the goals.
- *Mayoral leadership sets the tone.* Awareness of the SDGs in the USA remains fairly low. Mayor Garcetti's full-throated and public support for taking on the SDGs, and the ability of him and other senior city officials to articulate

the value proposition for Los Angeles, reinforces the value of the commitment to staff, residents, and external partners. It positions the city to take advantage of the mobilizing effect that can result from credible pursuit of outcome-based, time-bound goals.

3.6 Recommendations: Maximizing Value from the SDGs

The proposed LA SDG framework is holistic, ambitious, reflective of the city's values and priorities, and serious about focusing the city's attention on its most vulnerable populations and communities. The extent to which it will receive special attention, or be used by the mayor and senior officials as a tool or guidance for budget and policy decisions, remains unclear.

The city has opted not to create a special high-level SDG committee or internal task force. Its internal organization against the SDGs resembles a hub-and-spoke configuration, through which its chief SDG coordinator engages policy owners as appropriate throughout the city government.

It seems unlikely that the city will seek to create a comprehensive SDG-specific strategy to accompany a finalized framework. The proposed framework already contains and affirms many of the mayor's priorities, articulated through other processes and policies. Indeed, one might view all of Mayor Garcetti's executive directives, taken in total and combined, as the core of the city's SDG strategy.

Yet moving from alignment against the SDGs to analysis and action can create significant value for the city.

- *Analyze policy gaps and opportunities*:

Deeper analysis of the localized framework, by using evidence to map past trends and develop future scenarios, can identify areas where progress is likely to be insufficient, key challenges are going unnoticed, or opportunities for scaling high-impact interventions are hidden. Work undertaken in late 2018 by faculty and students from USC's Institute on Inequalities in Global Health to view the city's approach to homelessness through the lens of human rights provides a promising example. Universities and community-based organizations might also work with the appropriate city staff to use the multi-disciplinary aspects of the SDGs to surface new perspectives and develop integrated initiatives to advance progress on clustered issues, such as those related to homelessness.

The city presented a Voluntary Local Review (VLR) in 2019. A VLR is a report, notionally to the UN, of a city's specific contributions to the SDGs. Pioneered by New York City in 2018,[14] the format is based on the approach taken by countries' official reports on their SDG progress, presenting Voluntary National Reviews (VNRs) at the UN. The process of preparation provides an immediate opportunity to stimulate and incorporate such analysis. Publishing and publicizing the localized framework, once finalized, offer opportunities to community organizations and universities to undertake outside research relevant to city priorities.

- *Develop a platform for coordinated governance*:

While city government leadership will be instrumental, achieving the LA SDGs will depend upon strong shared city governance, with multiple segments of Los Angeles society contributing. Global experience with the MDGs and SDGs has demonstrated that specific, time-bound targets, with the right political attention and accountability, can have a mobilizing effect with businesses, investors, universities, civil society, and faith-based organizations. Los Angeles might explore models or platforms to enable public-private governance that generates and elevates multi-stakeholder efforts to advance specific priorities or the agenda overall. Hawai'i Green

[14] New York's Office for International Affairs (2018, July). *Voluntary Local Review: New York City's Implementation of the 2030 Agenda for Sustainable Development.* Retrieved from https://www1.nyc.gov/assets/international/downloads/pdf/NYC_VLR_2018_FINAL.pdf

Growth, for example, manages a public dashboard that measures Hawaii's progress and facilitates major public-private partnerships that contribute toward the state's SDG-aligned goals.

An immediate opportunity is to develop a platform for community organizations and citizens to be engaged in providing indicators and data relevant to the proposed LA SDGs. In creating its public dashboard, the city is adapting the open-source platform originally developed by the US chief statistician's office as the US reporting portal and could explore ways to integrate third-party community-level data. Another effort could engage the creative community in Los Angeles to develop storytelling and communications to reinforce and raise awareness of the city's commitment. Partners such as the Hilton Foundation and Occidental College can also act as champions to engage their networks and encourage collaborative action.

- *Use key SDG targets as a common denominator among different levels of government*:

In related fashion, the government of the city of Los Angeles does not have the statutory authority to achieve all the aspirations outlined in the proposed LA SDG framework. The proposed benchmarks and targets clarify the city's aspirations and can provide the basis for exploring coordinated action, or at least coordinated measurement, on select priorities to leverage respective authorities and resources among city, county, and state or federal government. The recent emerging cooperation between the city and county governments in reducing homelessness serves as a model. An executive directive by the mayor or a city council ordinance adopting the LA SDG framework would add credibility and weight to such efforts.

- *Explore new financing opportunities*:

Financial institutions, money managers, investors, and pension funds are exploring how the SDGs, as a globally vetted and agreed-upon framework, provide a standard framework for analyzing environmental, social, and governance factors. Firms like PIMCO are looking to structure SDG-specific product offerings, and the ratings agency Moody's recently published an assessment of the impacts of a Norwegian municipality's commitment to the SDGs on its future capital spending and borrowing.

- *Provide opportunities for engagement by residents*:

While overall awareness of the SDGs is low, support for the SDGs by the general public, once educated, is consistently positive. A recent poll by the UN Foundation found significant resonance among millennials. Other cities and municipalities have found the SDGs to be a compelling motivator for citizens and local groups to contribute toward the city's well-being.

The commitment made by Mayor Garcetti places Los Angeles in a leadership role among US cities taking on the SDGs. It seems unlikely that the proposed LA SDG framework, once finalized, will constitute the singular strategy used by the mayor and city council to define LA's future. Nevertheless, the city's investment in localizing the SDGs provides a comprehensive basis, one that is data-driven and outcome-focused, that can be a tool for enhancing and expanding solutions to improve the city's well-being. The challenge will be to take maximum advantage.

Tony Pipa is a senior fellow in Global Economy and Development at the Brookings Institution, where he researches place-based policies to improve social progress in the USA and globally, including use of the Sustainable Development Goals (SDGs) at the local level. He also studies the future of US multilateral aid and the applicability of lessons from international development to improving rural development in the USA. During the Obama administration, he served as chief strategy officer and held several other senior policy positions at USAID. He served as US Special Coordinator for the Post-2015 Agenda at the Department of State, leading the US delegation at the UN to negotiate and adopt the SDGs. Prior to his government service, he directed the NGO Leaders Forum at Harvard University and was the founding CEO of the Warner Foundation. He serves on the board of directors of StriveTogether and the Advisory Council of the Center for Disaster Philanthropy. He attended Stanford University, graduated from Duke University, and earned a Master of Public Administration at the Harvard Kennedy School.

4 The Idea of Sustainability to Evaluate Growth and Development in the Houston Region

David B. Abraham

4.1 Introduction

The Sustainable Development Goals (SDGs) were developed for national-level reporting and with performance target differences between countries in mind. Performance targets in the SDGs also reflect differences between islands and countries, large and small countries, and countries with a range of development standards and capacities. What is not properly reflected in the SDGs are local-level performance targets. At the local level, there are vastly many more differences between places than at the national level. At the local level, in the USA, there is a great deal of political autonomy and hence ability to impact sustainability in ways that may not be possible at the national level. Since local political autonomy is so strong in the USA, it is important to understand differences in development performance between cities. This is essential to be able to identify appropriate policies and programs to enhance sustainability both locally and at the national level. In this chapter we show how the Houston region compares to other regions in the country. Properly understanding these differences will enhance our ability to set appropriate targets, policies, and programs to enhance sustainability across the Houston region.

In this chapter, we review major sustainability challenges, opportunities, and constraints for the Houston metropolitan region. We first cite past and present major efforts at developing sustainability programs in the region. Then we highlight unique challenges faced in the Houston region through a comparison to the top 100 most populous metros in the USA.

4.2 From the Past to the Present of Sustainability Efforts in the Houston Region

Over the last 45 years, there have been at least four (4) major efforts at developing sustainability programs in the Houston region. The first major effort occurred in 1974 when tycoon and engineer George Mitchell[1] convened a group of renowned business leaders, educators, and futurists, including Dennis Meadows[2] in 1974, to

D. B. Abraham (✉)
Rice University, Houston, TX, USA
e-mail: DAbraham@Rice.edu

[1] George Mitchell is considered "the Father of Shale Gas." His company, Mitchell Energy and Development Corporation, was sold to Devon Energy in 2002 for $3.5 billion. Between the 1980s and 1990s, Mitchell's company pioneered horizontal drilling technology and combined this with hydraulic fracturing of rock. This breakthrough called "fracking" made it economically possible to extract natural gas from shale rock.

[2] Dennis and Donella Meadows published *The Limits to Growth* in 1972. The book, which sold 20 million copies, was the result of a study commissioned by the Club of Rome in 1970, to study "the predicament of mankind." Meadows and his 16-member team designed a macroeconomic model to analyze connections between population growth, food production, use of natural resources, industrial production, and pollution.

D. B. Abraham, S. D. Iyer (eds.), *Promoting the Sustainable Development Goals in North American Cities*, Sustainable Development Goals Series, https://doi.org/10.1007/978-3-030-59173-1_4

develop a 10-year program of action to identify alternatives to the *then-present-and-projected nature of growth and development*. The first of four conferences was convened in the Houston region in 1975. The conference featured prominent experts in development services such as Dennis and Donella Meadows, Lester Brown, Herman Kahn, Ian McHarg, and Herman Daly. Donella Meadows authored a paper for the conferences titled "Equity, the Free Market, and the Sustainable State." The second conference was convened in 1977, co-sponsored by the Club of Rome. The conference was titled *The Nature of Growth in Equitable and Sustainable Societies*. The third conference in 1979 was titled *Management of Sustainable Growth*. Finally, the fourth conference was convened in 1982 titled *Role of the Private Sector* (Schmandt 2010).

The second and third major regional efforts of note were both initiated in 2010. The Houston regional council of governments received funding from President Barack Obama's national sustainability enhancement program, *Partnership for Sustainable Communities*. The grant was awarded to develop a regional plan for sustainability (HGAC 2011). That same year the City of Houston, the most populous city in the Houston metro region, created a position for a sustainability director.

The fourth major effort actually started prior to efforts two and three cited above, but is presented as number four to contextualize this author's affiliation and the study that is the focus of the rest of this chapter. In 2002, Rice University launched the Shell Center for Sustainability, to foster and develop interdisciplinary research partnerships to evaluate sustainable development in the region. Over the course of 15 years, the research think tank convened business, government, and civil stakeholders authoring numerous studies and reports on sustainability in the region (SCS 2017). This author and the Rice University, Sustainability Solutions Lab, are both birthed from that effort. Beginning in 2011, we published reports comparing the City of Houston to peer cities in the USA on key sustainability indicators (Blackburn 2011). Several sustainability indicators reports and articles were published, analyzing the City from varying units of geography (King 2012, 2013, 2014, 2015, 2016, 2017). In 2018 our research team developed partnerships with the Center of Sustainability and Resilience at the University of Houston and with the International Institute for Sustainable Development, to expand our research and reporting capabilities to address the Sustainable Development Goals (SDGs) for Houston (Sustainability Solutions Lab 2018; Abraham et al. 2019). The research partnership team's goal is to convene industry and the civil sector in the region, to collaboratively develop policies and programs for the Houston region to meet the SDG 2030 goals.

4.3 Sustainability and the State of Growth and Development in the Houston Region

The Houston metropolitan region in Texas is composed of nine counties: Austin, Brazoria, Chambers, Fort Bend, Galveston, Harris, Liberty, Montgomery, and Waller counties (Fig. 4.1). The total population was 6,482,592 in 2016 and the total land area is 8261 square miles (ACS_16). Houston is the sixth most populous metro in the country behind New York, NY; Los Angeles, CA; Chicago, IL; Dallas, TX; and New York, NJ. Since Houston has relatively the same land size as the New York metro, it is approximately half as dense as New York (1613/sqml in New York compared to 785/sqml in Houston). From a sustainability perspective, this immediately signals issues for Houston, stemming from sprawl. Issues such as the higher infrastructure management burden for Houston and the emissions generated from moving people and goods across a much larger geography are all issues Houston has to manage compared to New York (Table 4.1).

The Houston metro is a large geographic area with 1000s of individual and overlapping governance bodies. The primary sustainability challenge and major constraint for the Houston metro region is one of *governance*. Houston is also widely considered to be the least planned, least

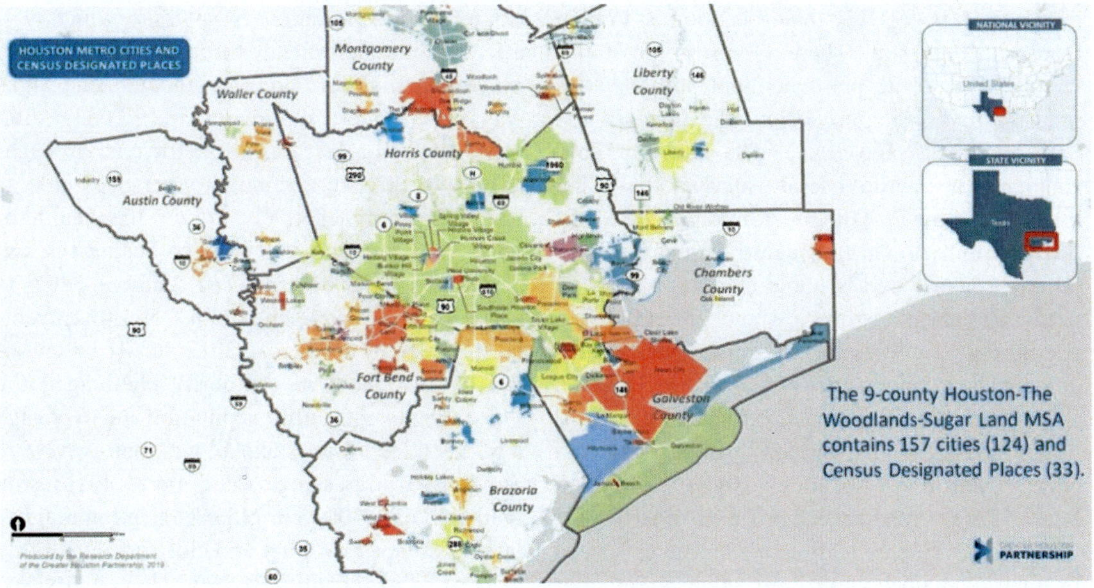

Fig. 4.1 Houston-Galveston metropolitan region

Table 4.1 Top 20 MSAs ranked by 2016 population

	Top 20 MSAs ranked by 2016 population	Population	Population density (Pop/SqMI)	Area (SqMI)
1	New York-Newark-Jersey City, New York	13,380,318	1613.14	8294.58
2	Los Angeles-Long Beach-Anaheim, California	13,189,366	2720.13	4848.81
3	Chicago-Naperville-Elgin, Illinois	8,656,303	1202.81	7196.75
4	Dallas-Fort Worth-Arlington, Texas	6,957,123	749.68	9280.14
5	New York-Newark-Jersey City, New Jersey	6,594,915	795.09	8294.58
6	**Houston-The Woodlands-Sugar Land, Texas**	**6,482,592**	**784.74**	**8260.82**
7	Miami-Fort Lauderdale-West Palm Beach, Florida	5,926,955	1167.76	5075.51
8	Atlanta-Sandy Springs-Roswell, Georgia	5,612,777	646.54	8681.32
9	San Francisco-Oakland-Hayward, California	4,577,530	1847.45	2477.75
10	Phoenix-Mesa-Scottsdale, Arizona	4,486,153	308.01	14,565.04
11	Riverside-San Bernardino-Ontario, California	4,430,646	162.51	27,263.57
12	Boston-Cambridge-Newton, Massachusetts	4,302,566	1234.29	3485.86
13	Detroit-Warren-Dearborn, Michigan	4,296,731	1104.81	3889.10
14	Philadelphia-Camden-Wilmington, Pennsylvania	4,076,378	885.66	4602.63
15	Seattle-Tacoma-Bellevue, Washington	3,671,095	625.33	5870.66
16	Minneapolis-St. Paul-Bloomington, Minnesota	3,360,829	440.07	7637.08
17	San Diego-Carlsbad, California	3,253,356	773.39	4206.64
18	Tampa-St. Petersburg-Clearwater, Florida	2,927,714	1164.30	2514.58
19	Washington-Arlington-Alexandria, Virginia	2,884,799	461.81	6246.67
20	Baltimore-Columbia-Towson, Maryland	2,780,873	1068.94	2601.52

Source: ACS_5Yr_16

regulated urban area in the USA (Pendall et al. 2006). So, to identify collaborative efficiencies within and between the multiple government agencies is to evaluate the contemporary nature of governance in one of the more neoliberal regions in the USA.

A review of the map of the Port of Houston region shows how geographically fragmented governance is for this major economic hub in Houston (Fig. 4.2). There is no single entity which governs the petrochemical and logistics industry in this region, below the level of the State of Texas (Table 4.2). The Houston Ship Channel is a 52-mile federal waterway that is the backbone of the Port of Houston region. The Port of Houston region is considered the energy capital of the world and is home to a wide array of petrochemical industries, oil and gas refineries, and storage facilities for shipping and logistics companies. Economic activity along the Houston Ship Channel was estimated at nearly $802 billion, 3.2 million jobs, and $38 billion of tax revenue (Martin Associates 2018). Although laissez-faire governance can be an enabler for business development in some instances, it can become a major liability in other areas. For example, transportation is an issue which will affect business efficiency throughout this region. Since personal cars are sharing the same highways as commercial vehicles, traffic congestion increases will affect business efficiency in negative ways. From a sustainability perspective, traffic congestion negatively affects the health of employees, economic output of businesses and employees, and overall environment, due to higher air pollution generation from idling vehicles. The State of Texas' 2012 sunset review of the Port of Houston Authority challenged the Port to remake its public reputation and to engage a wide collection of stakeholders than they were previously addressing. "The Port of Houston Authority is a 100-year old organization that has largely escaped scrutiny and close accountability because of its location between state, county, and

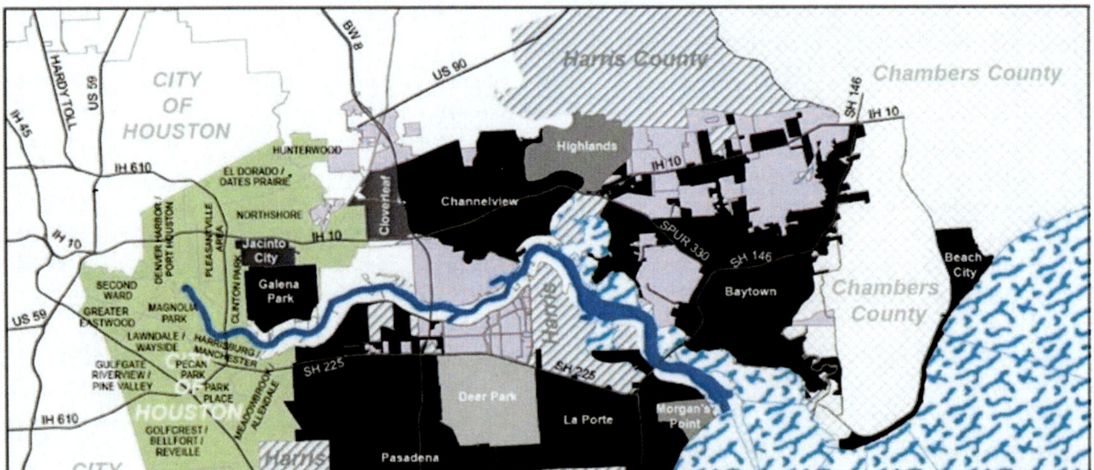

Fig. 4.2 Port of Houston region

Table 4.2 Governance bodies in the Port of Houston industrial region

Port of Houston governance entities (partial list)
1. Counties – Harris County, Galveston County, Chambers County
2. Cities – Houston, Galena Park, Pasadena, Shoreacres, La Porte, Morgan's Point, Seabrook, Deer Park, Galveston Bay, Channelview, Cloverleaf, Highlands, Baytown, Beach City, South Houston, Jacinto City, Taylor Lake Village, El Lago, Sheldon, Barrett
3. Industrial Districts: Houston IND Dist, Pasadena IND Dist, Deer Park IND Dist, La Porte IND Dist, Baytown, IND Dist
4. Municipal Utility Districts (MUDs) – COH MUDs, Harris County Municipal Utility Districts (MUDs), Fresh Water Supply Districts (FWSDs), Water Control and Improvement Districts (WCIDs)

city governments, away from their direct oversight or control" (Final Results of Sunset Reviews 2012–2013). Part of the outcome of the sunset review by the state was to replace all board members of the Port of Houston, except for its chairperson. The state also appointed an additional seat on the commission. The purpose of this additional seat was primarily to manage the multimodal implications of the regional activity across the Port of Houston.

These sustainability issues stemming from fragmented governance are also highlighted in other sectors in the Houston region, such as k-12 education, healthcare accessibility, food accessibility, and natural hazard resilience. In the k-12 education sector, the Houston Independent School District (HISD) is currently slated for state takeover. The Texas Education Agency (TEA) commissioner has announced plans on replacing the school board based on school performance and board governance issues (Texas Education Agency 2019). Related issues in the k-12 sectors include preparing students to fill the growing workforce skills gap identified by Houston area companies. In 2014, the Greater Houston Partnership initiated an industry-led task force and subsequently a program called UpSkill Houston to develop intervention strategies to bridge the skills and training gap (Greater Houston Partnership 2014). In the natural hazard resilience sector, the governor of Texas recently withheld $4.3 billion in federal mitigation funding instead of direct appropriations to the City of Houston for local management (Wallace 2019). Food and healthcare access have not yet had state intervention. Part of the reason may be that these two areas represent sectors that are wholly privatized. However, a review of the region's performance in these sectors will highlight additional growing issues. Houston is ranked 92nd out of the top 100 metros in the country with regard to food insecurity. Houston is ranked 97th on access to healthcare and 89th on access to primary care physicians. These last two point to clear disparities between our rank and performance in comparison to other areas in the country and the fact that the Texas Medical Center, located centrally in Houston, is the largest medical center in the world.

4.4 Houston's Development Performance on the SDGs

Across the 15 SDGs and 44 metrics used in this study, the Houston metro did not compare very well when compared to the top 100 metros in the country. Houston ranked in the bottom half of all metros used in this study on 31 of the 44 measures. Houston ranked in the top 20 metros for two measures. On SDG 11 Sustainable Cities and Communities, Houston ranked 13th for housing affordability; and on SDG 5 Gender Equality, Houston ranked 15th for number of women-owned businesses (Table 4.3).

The Houston metro has another 10 years to reach the 2030 SDGs. Beginning in 2020, our research team will engage major industries and local governments in the region to develop a 10-year plan of action to become more sustainable by 2030. The following section presents Houston's performance on each of the 15 SDGs. We highlight the leading metro for each of the

Table 4.3 Houston metro performance on SDGs compared to the top 100 most populous metros in the USA

Houston metro performance on SDGs compared to the top 100 most populous metros in the USA

SDGS		Data	100 metros N	100 metros minimum	100 metros maximum	100 metros mean	Houston metro	Houston metro rank
1 NO POVERTY	**Pop below poverty**	%	100	8.4	32.8	14.5	**15.3**	**66**
	Income	$/capita	99	24,805.0	106,666.0	48,758.9	51,900.4	74
	Children x2 below poverty line	%	99	25.2	69.7	41.0	40.7	51
2 ZERO HUNGER	**Food insecurity**	%	100	7.9	20.6	13.4	**16.3**	**92**
	Pop obese	%	100	19.1	34.4	27.9	26.9	37
3 GOOD HEALTH AND WELL-BEING	**Infant mortality**	1/1000	93	3.4	12.1	6.0	5.4	39
	Low birth weight	%	93	5.5	11.2	8.0	**8.6**	**67**
	STDs	1/100000	99	195.8	1051.4	624.0	**655.6**	**58**
	Diabetes	1/1000	100	55.9	143.0	98.2	87.8	23
	Heart attack deaths	1/100000	100	59.1	257.8	166.9	**169.7**	**49**
	Traffic accident deaths	1/100000	100	5.1	17.8	10.1	**11.5**	**67**
	Health insurance	%	100	67.1	96.6	89.1	**80.1**	**97**
	Primary care physicians	1/100000	100	43.9	115.7	78.6	**59.9**	**89**
4 QUALITY EDUCATION	**School enrollment**	%	100	89.7	96.2	94.1	94.3	40
	Population without high school degree	%	100	5.2	36.7	12.1	**17.6**	**92**
	School enrollment 3–4 years	%	100	31.2	70.5	47.3	**43.8**	**63**
	Degree undergraduate	%	100	15.7	49.4	31.5	**31.3**	**47**
5 GENDER EQUALITY	**Earnings gap**	%	100	16.1	56.0	28.1	27.8	54
	Rapes	1/100000	94	7.2	123.8	42.7	39.9	50
	Women businesses	%	100	29.9	46.3	36.2	39.3	15
6 CLEAN WATER AND SANITATION	**Deficit water stress**	Index	98	0.0	79.9	6.5	**8.1**	**75**
7 AFFORDABLE AND CLEAN ENERGY	**Low carbon energy generated**	%	100	3.8	85.9	35.1	**22.8**	**65**

(continued)

Table 4.3 (continued)

Houston metro performance on SDGs compared to the top 100 most populous metros in the USA

8 DECENT WORK AND ECONOMIC GROWTH	**GDP growth**	%	100	−0.7	6.6	2.0	2.9	23
	Jobs STEM	%	100	5.9	22.1	11.9	11.9	66
	Unemployment	%	100	2.3	9.3	4.3	**5.1**	**85**
	Disconnected youth not working	%	94	7.7	21.6	13.5	**14.2**	**56**
9 INDUSTRY, INNOVATION AND INFRASTRUCTURE	**Utility patent grants**	1/1000	100	0.3	53.3	4.6	4.3	29
	Broadband	%	100	64.6	90.8	82.6	83.3	63
10 REDUCED INEQUALITIES	**Gini coefficient**	Index	100	0.4	0.5	0.5	0.5	90
	Racial segregation	Index	97	21.9	81.5	55.1	**61.4**	**63**
	Upward mobility	Index	100	33.7	49.2	40.9	42.9	27
11 SUSTAINABLE CITIES AND COMMUNITIES	**Housing affordability**	$	100	2.3	7.9	3.6	2.6	13
	Pop overcrowded housing	%	100	0.7	12.8	3.1	**5**	**86**
	Sustainable transportation	%	100	1.4	37.6	6.0	**4**	**58**
	Sprawl index	Index	100	1.4	37.6	6.0	**4**	**58**
	Access to parks	%	100	8.0	81.5	41.8	**26**	**79**
	PM 2.5 average	Micrograms per cubic meter	99	6.5	14.8	10.0	**11.7**	**85**
	Ozone levels	ppm	81	0.048	0.106	0.071	**0.077**	**84**
12 RESPONSIBLE CONSUMPTION AND PRODUCTION	**Toxic releases**	lbs per square mile	100	3.5	33,728.9	1432.9	**9,601.7**	**97**
13 CLIMATE ACTION	**Household carbon footprint**	TCO2e per capita	100	12.4	51.4	22.9	**33.7**	**89**
15 LIFE ON LAND	**Open space**	sq-meters per capita	87	14.9	226.0	84.5	No data	No data
	Brownfield and Superfunds	Sites per square mile	100	0.0	0.5	0.1	0.04	49
16 PEACE, JUSTICE AND STRONG INSTITUTIONS	**Violent crimes**	1/100000	92	8.5	1262.3	425.3	**578.2**	**75**
	Firearms deaths	1/100000	100	2.1	25.6	10.3	**11.6**	**66**

Data Source: (Espey et al. 2018)

measures in the study and the 2030 SDG target all metros should be aiming policies and programs to meet.

4.4.1 SDG1: No Poverty

Poverty rates in Houston are increasing (GHP 2019). Here we measure (1) the number of people in poverty and (2) average incomes and (3) we focus on children in poverty. In 2010, the City of Houston had a higher poverty rate (22.8%) than its county seat Harris County and then in turn the State of Texas (16.8% for both). The percent of people below the poverty line in the USA was 13.8% (US Census Bureau, 2010). This range, from the mostly urban city of Houston to the mostly rural State of Texas, suggests that interventions in the region need to have sensitivity between different types of geographies such as urban cities vs suburban and rural cities. The City of Houston is projected to have 24% of its population in poverty by 2030 unless major interventions are enacted (King 2012).

SDG1		Data	N	Minimum	Maximum	Mean	Houston
	Pop below poverty	%	100	8.4	32.8	14.5	**15.3**
	Income	$/capita	99	24,805.0	106,666.0	48,758.9	51,900.4
	Children x2 below poverty line	%	99	25.2	69.7	41.0	40.7

Source: (Espey et al. 2018) Based on 2016 data

For percentage of the population below poverty, the best metro target is projected to be 4.2%. This is based on the SDG1 to half the poverty level by 2030. The highest performing metro was the Washington, DC – Arlington metro with 8.4%. The Houston metro value was 15.3% of the population below the poverty level. This performance needs to be strengthened since it is more than three times the projected 2030 SDG target. Houston ranks 66th among the 100 most populated metros in the country on this indicator.

For personal income the 2030 SDG target is projected at $83,025.41 per capita. The top 5 in descending order were Bridgeport-Stamford-Norwalk, San Jose, San Francisco, Boston followed by New York. The Houston metro value was $51,900.40 per capita in 2016. This is significantly above the average for the country but below the 2030 target which is set at $83,025.41. The highest performing metro area in the country was Bridgeport-Stamford with a $106,666 per capita income in 2016. Houston ranks 74th in the country on this indicator.

The percent of children that are living more than two times below the poverty line was 40% in the Houston metro. The target for this indicator is 12.5%, which is based on the SDG1 to halve the number of children living two times below the poverty level by 2030. This indicator needs to be strengthened in the metro region. The best performing metro on this indicator was 25.2% in the Albany metro. Houston ranks 51th in the country on this indicator.

4.4.2 SDG2: Zero Hunger

SDG2		Data	N	Minimum	Maximum	Mean	Houston
	Food insecurity	%	100	7.9	20.6	13.4	**16.3**
	Pop obese	%	100	19.1	34.4	27.9	26.9

Source: (Espey et al. 2018). Food insecurity 2014 data. Population obese 2013 data

The target for the percentage of the population to be food insecure is set at 0. The best performing metro was the Washington DC metro with 8% in 2014. The Houston metro value was 16.3%. The performance in Houston needs to be strengthened to reduce the number of food-insecure people. Houston ranks 92nd in the country on this indicator.

The target for the percentage of people with obesity is set at 2.8% based on the Global SDG Index. Obesity is here defined as having a body mass index (BMI) of 30 or higher. The Houston metro score on this indicator was 26.9%. Although this value is lower than the average metro value of 27.9%, it is still much higher than the projected target of 2.8% by 2030. The metro with the best performance in 2013 was San Diego. Houston ranks 37th in the country on this indicator.

4.4.3 SDG3: Good Health and Well-Being

SDG3		Data	N	Minimum	Maximum	Mean	Houston
	Infant mortality	1/1000	93	3.4	12.1	6.0	5.4
	Low birth weight	%	93	5.5	11.2	8.0	**8.6**
	STDs	1/100000	99	195.8	1051.4	624.0	**655.6**
	Diabetes	1/1000	100	55.9	143.0	98.2	87.8
	Heart attack deaths	1/100000	100	59.1	257.8	166.9	**169.7**
	Traffic accident deaths	1/100000	100	5.1	17.8	10.1	**11.5**
	Health insurance	%	100	67.1	96.6	89.1	**80.1**
	Primary care physicians	1/100000	100	43.9	115.7	78.6	**59.9**

Source: (Espey et al. 2018). Infant mortality 2015 data. Low birth Weight 2016 data. STDs 2015 data. Diabetes 2013 data. Heart attack deaths 2015 data. Traffic deaths 2008–2014 data. Health Insurance 2016 data. PCPs 2015 data

The number of infant deaths per 1000 live births is the metric used to define the indicator infant mortality. The global target is set at 2.8, which is based on the average of the top 5 OECD countries. The best performing metro in the county was San Jose with 3.4%. Houston scored 5.4% on this indicator. This is higher than the average for the country but below the target set for 2030. Houston ranks 39th in the country on this indicator.

The percent of infants born with a low birth weight is defined as those born weighing less than 2500 grams. The target is set at 4.3%, which is based on the average of the top 5 OECD countries on this measure. In the USA, the Oxnard metro in California was the best performing metro on this measure with 5.5%. In the Houston metro, 8.6% of every 1000 births are low birth weight. This is an area that we need to strengthen in our region. The average among the 100 metros in this study was 8% of all babies. Houston ranks 67th in the country on this indicator.

The STD rate is tracked at the number of positive cases per 100,000 persons. STDs tracked for this measure are syphilis, chlamydia, and gonorrhea cases. The Houston rate was 656 positive persons out of every 100,000 in 2015. This is higher than the national average of 624/100,000 persons. It is also significantly higher than the global target of 262/100,000 persons. The best performing metro on this indicator was the Provo metro in Utah, with a rate of 196 positive STD cases out of every 100,000 persons. Houston ranks 58th in the country on this indicator.

The diabetes measure is an aggregate of all types of diabetes cases per every 1000 people. The 2030 target is set at 65 persons. In the Houston metro, 88 people out of every 1000 have some form of diabetes. This is below the national average of 98/1000 but higher than the 2030 target. The best performing metro was the Provo metro in Utah, with 56 people out of every 1000 with some form of diabetes. Houston ranks 23th in the country on this indicator.

Heart attack deaths are measured by the number of deaths for people over the age of 35 per 100,000 people. In the Houston metro, approximately 170 people out of every 100,000 died of heart attacks in 2015. This is higher than the 2030 target of 31 and higher than the national average of 167 persons. The metro with the best performance on this indicator was the Cape Coral metro in Florida, with 59 deaths per every 100,000 people. Houston ranks 49th in the country on this indicator.

The 2030 target for traffic accident deaths is 3 persons per every 100,000 persons. The Houston metro had 11.5 deaths for every 100,000. The average for the country was 10 for every 100,000. The best performing metro in this area was the Boston metro, with 5 deaths out of every 100,000 persons. Houston ranks 67th in the country on this indicator.

The percentage of persons with health insurance in the Houston metro was 80% in 2016. This is lower than the national average of 89% and below the 2030 target of 100%. The highest performing metros in the country were Boston, Springfield, and Worcester all in Massachusetts, with approximately 97% of persons with health insurance. Houston ranks 97th in the country on this indicator.

The number of primary care physicians (PCPs) per every 100,000 persons was approximately 60% in the Houston metro in 2015. The Houston metro has the largest medical center in the world. The average for the country was approximately 79%, and the highest performing metro was San Francisco, with approximately 116 PCPs per every 100,000 people. This number surpasses the global 2030 target of 110. Houston ranks 89th in the country on this indicator.

4.4.4 SDG4: Quality Education

SDG4		Data	N	Minimum	Maximum	Mean	Houston
	School enrollment	%	100	89.7	96.2	94.1	94.3
	Population without high school degree	%	100	5.2	36.7	12.1	**17.6**
	School enrollment 3–4 years	%	100	31.2	70.5	47.3	**43.8**
	Degree undergraduate	%	100	15.7	49.4	31.5	**31.3**

Source: (Espey et al. 2018). School enrollment 2016 data. Population without high school degree 2016 data. School enrollment 3–4 years of age 2016 data. Undergraduate degree 2016 data

Higher levels of education directly produce healthier behaviors and better jobs and income outcomes (Sanborn 2012). Education attainment is critical to economic, civil, and personal health viability. School enrollment (5–19 years) was 94.3% for the Houston metro in 2016. This is slightly above the national average of 94.1% but below the aspirational 2030 target of 100%. The San Jose and San Francisco metros in California and the Madison, Wisconsin metro all tied for the highest score on this indicator with 96% of 5–19-year-olds enrolled in school. Houston ranks 40th in the country on this indicator.

High school dropouts in the Houston metro were 17.6% in 2016. This is above the national average of 12.1% and above the 2030 SDG target of universal secondary education (0% dropouts). The Madison, Wisconsin metro had the lowest score on this indicator with a high school dropout rate of 5.2%. Houston ranks 92nd in the country on this indicator.

Preschool enrollment was 43.8% in Houston in 2016. This is below the national average of 47.3% and below the aspirational 2030 SDG target of 100% preschool enrollment. The Bridgeport, Connecticut metro had the highest preschool enrollment with 70.5 kids aged 3–4 in school. Houston ranks 63rd in the country on this indicator.

The Houston metro had 31.3 persons with undergraduate degrees or higher in 2016. This is very slightly below the national average of 31.5% and below the 2030 SDG target of 48.7%. The metro with the highest undergraduate degree rate or higher was the Washington DC metro, with 49.4%, which is higher that the global target for 2030 of 48.7%. Houston ranks 47th in the country on this indicator.

4.4.5 SDG5: Gender Equality

SDG5		Data	N	Minimum	Maximum	Mean	Houston
	Earnings gap	%	100	16.1	56.0	28.1	27.8
	Rapes	1/100000	94	7.2	123.8	42.7	39.9
	Women businesses	%	100	29.9	46.3	36.2	39.3

Source: (Espey et al. 2018). Earnings gap 2016 data. Rapes 2016 data. Women owned businesses 2012 data

The gender wage gap is defined as the percentage difference between the median wage of men and women. The Houston metro had a 27.8% wage gap in 2016. This is a better performance than the national average of 28.1%. The aspirational 2030 target is set at a 0% wage gap. The Las Vegas, Nevada metro had the best performance in the country in 2016 with a 16.1% wage gap. Houston ranks 54th in the country on this indicator.

The reported rapes per 100,000 women was 39.9 in 2016. This is a lower number than the national average of 42.7. The best performing metro on this indicator was Kansas City with 7.2 rapes per every 100,000 people. Houston ranks 50th in the country on this indicator.

The percentage of women-owned businesses in the Houston metro was 39.3% in 2012. This is above the national average of 36.2%. The Memphis metro had the highest percentage of women-owned businesses with 46.3%. The aspirational 2030 target is 50% of all businesses to be owned by women. Houston ranks 15th in the country on this indicator.

4.4.6 SDG6: Clean Water and Sanitation

SDG6		Data	N	Minimum	Maximum	Mean	Houston
	Deficit water stress	Index	98	0.0	79.9	6.5	**8.1**

Source: (Espey et al. 2018). Deficit Water Stress Index 2009 data

The Water Stress Index value for the Houston metro was 8.1 in 2009. This index follows the amount of average annual rainfall needed to remove water stress from an area. Houston does not perform as well as the average metro in the country which has a stress index of only 6.5. The best performing metros for the country were Manchester, New Hampshire; Palm Bay, Florida; Charleston, South Carolina; and Seattle, Washington, each with a water stress index of 0. Houston ranks 75th in the country on this indicator.

4.4.7 SDG7: Affordable and Clean Energy

SDG7		Data	N	Minimum	Maximum	Mean	Houston
	Low carbon energy generated	%	100	3.8	85.9	35.1	**22.8**

Source: (Espey et al. 2018). Low carbon energy generated 2016 data

The Houston metro reported a 22.8% generated energy from low carbon sources in 2016. Low carbon is defined here as wind, solar, geothermal, biomass, hydroelectric, and nuclear. This was below the national average of 35.1%. The 2030 SDG target for low carbon energy generated is 78.7%. The Seattle and Spokane metros both exceed the 2030 target with an 85.9% low carbon generated energy mix. Houston ranks 65th in the country on this indicator.

4.4.8 SDG8: Decent Work and Economic Growth

SDG8		Data	N	Minimum	Maximum	Mean	Houston
	GDP growth	%	100	−0.7	6.6	2.0	2.9
	Jobs STEM	%	100	5.9	22.1	11.9	11.9
	Unemployment	%	100	2.3	9.3	4.3	**5.1**
	Disconnected youth not working	%	94	7.7	21.6	13.5	**14.2**

Source: (Espey et al. 2018). GDP Growth 2011–2016 data. Stem Jobs 2016 data. Unemployment 2017 data. Disconnected youth 2015 data

The size of the workforce is often seen as a good economic indicator because it points to the human resource capacity of the area to drive the economy (Kotkin, 2007). Disagreements however have surfaced over whether emphasis should be placed on primarily developing the professional class of workers or the service workers as key engines to economic success in the twenty-first century (Florida, 2002; Kotkin, 2007). SDG8 does well to focus on growth of STEM jobs as an indicator of quality employment. Environmental considerations should be considered as well as to whether these companies produce a lot of waste or pollution or whether they are more environmentally responsible companies (Wooster 2013).

GDP growth for the Houston metro was 2.9% in 2016. This was higher than the national average. The San Jose metro had the highest GDP growth rate of 6.6%. This was higher than the 2030 SDG target which is 5.8%. Houston ranks 23rd in the country on this indicator.

The percentage of STEM jobs for Houston was equal to the national average of 11.9%. STEM jobs are jobs in computers, science, engineering, healthcare, and technical occupations. The top performing metro in the country San Jose had 22.1% of jobs in STEM fields. Houston ranks 66th in the country on this indicator.

The unemployment rate for the Houston metro was 5.1% in 2017. This is higher than the national average of 4.3% and higher than the 2030 SDG target of 3.7%. The top performing metro in the country Urban Honolulu had a 2.3% unemployment rate. Houston ranks 85th in the country on this indicator.

The percentage of youth (ages 16–24) who are not in education, employment, or training (NEET) is considered the metric for disconnected youth. In the Houston metro, 14.2% of youth are in the NEET category. This is higher than the national average of 13.5% and higher than the 2030 SDG target of 8.3%. The metro with the lowest NEET rate was Omaha, with 7.7%. Houston ranks 56th in the country on this indicator.

4.4.9 SDG9: Industry, Innovation, and Infrastructure

SDG9		Data	N	Minimum	Maximum	Mean	Houston
	Utility patent grants	1/1000	100	0.3	53.3	4.6	**4.3**
	Broadband	%	100	64.6	90.8	82.6	83.3

Source: (Espey et al. 2018). Patent applications 2012–2015 data. Broadband connection 2016 data

The number of utility patent applications per 1000 workers in the Houston metro was 4.3 in 2015. This was below the national average of 4.6. The highest number of patent applications in the nation came out of San Jose, with 53.3 patent applications in 2015. The SDG 2030 goal is 22.8. Houston ranks 29th in the country on this indicator.

The Houston metro is ahead of the national average on broadband connectivity with 83.3% of households having access to a broadband connection. The average in the nation is 82.6% of households. The leading metros, San Jose, California and Colorado Springs, Colorado, had 91% of households connected to a broadband network. Houston ranks 63rd in the country on this indicator.

4.4.10 SDG10: Reduced Inequalities

SDG10		Data	N	Minimum	Maximum	Mean	Houston
	Gini coefficient	Index	100	0.39	0.54	0.5	0.5
	Racial segregation	Index	97	21.9	81.5	55.1	**61.4**
	Upward mobility	Index	100	33.7	49.2	40.9	42.9

Source: (Espey et al. 2018). Gini Coefficient 2016 data. Racial Segregation 2010 data. Upward mobility 2016 data

The Gini coefficient index measure for the Houston metro was 0.5 in 2016. This is a measure for the distribution of income inequality. The range among US metros was 0.39 in Ogden, Utah, which was the best performing metro, to 0.54 in the Bridgeport, Connecticut metro. The global SDG index target is 0.25 by 2030, which suggests that all metros still have the challenge of bringing the gap between income disparities in the USA. Houston ranks 90th in the country on this indicator.

The racial segregation index measures the degree to which African-Americans are distributed differently to which Caucasians across geographic census tracts. The best performing metro on this measure was Provo, Utah, with an index score of 21.9. The average segregation measure across all measures was 55.1 and the Houston metro scored 61.4. Houston ranks 63rd in the country on this indicator.

The upward mobility index average across all US metros was 40.9. The top performing metro was 49.2 in Provo, Utah. The Houston metro score was 42.9, which is higher than the national average of 40.9. Houston ranks 27th in the country on this indicator.

4.4.11 SDG11: Sustainable Cities and Communities

SDG11		Data	N	Minimum	Maximum	Mean	Houston
	Housing affordability	$	100	2.3	7.9	3.6	2.6
	Pop overcrowded housing	%	100	0.7	12.8	3.1	**5**
	Sustainable transportation	%	100	1.4	37.6	6.0	**4**
	Sprawl index	Index	100	1.4	37.6	6.0	**4**
	Access to parks	%	100	8.0	81.5	41.8	**26**
	PM 2.5 average	Micrograms per cubic meter	99	6.5	14.8	10.0	**11.7**
	Ozone levels	ppm	81	0.048	0.106	0.071	**0.077**

Source: (Espey et al. 2018). Housing affordability 2016 data. Overcrowded housing 2016 data. Sustainable transportation 2016 data. Sprawl index 2010 data. Access to parks 2015 data. PM2.5 2015 data. Ozone levels 2016 data

As populations increase, there is a current practice to spend more on highways for transportation efficiency. However, autos emit large CO levels and ozone precursors (VOCs and NOx) during fuel combustion, and there is also evaporation during operation and resting (EPA 2001; TRB 1995). Since there is a direct link between land use and transportation and vehicle exhaust produces precursors for ozone, therefore local land use plans should reduce transportation dependence to support efforts to lower ozone levels (SGN 2002). The challenge for municipalities is reducing air pollution and unhealthy days while reducing traffic congestion concurrently.

Housing affordability is measured here as the median property value divided by the median household income in the metro area. The best performing metro in the country was the McAllen, Texas metro, which scored 2.3 on this measure. The Houston metro scored 2.6 which was better than the national average of 3.6. The 2030 SDG target is set at 2.4, which means Houston is relatively close to meeting this goal. Houston ranks 13th in the country on this indicator.

The percentage of the population living in overcrowded housing was 5% in the Houston metro, which was higher than the national average of 3.1. Akron, Ohio was the best performing metro with only 0.7% of the population living in overcrowded housing. Houston ranks 86th in the country on this indicator.

The percentage of workers using sustainable transportation to get to work (transit, bicycles, walking) was 4% in the Houston metro. The best performing metro in the country was the New York metro area, with 37.6% of workers using sustainable transportation to get to work. The 2030 target is 22.7% of the population using sustainable transportation. Houston ranks 58th in the country on this indicator.

The sprawl index is a composite of measures including residential and employment density; neighborhood mix of homes, jobs, and services; strength of activity centers and downtowns; and accessibility of the street network. The range among US metros was from the lowest performing Jackson, Mississippi metro with a 1.4 score on the index to the highest performer New York metro with a 37.6 score on the index. The average among US metros was 6 and Houston is behind the average with a score of only 4 on the sprawl index. Houston ranks 58th in the country on this indicator.

Parks access is measured by the percentage of people living within 15 min from a public park. The average in the country among US metros was 41.8%. The top performing metro was San Francisco, with 81.5% of the population living within 15 min to a public park. In the Houston metro, 26% of the population live within 15 min to a public park. The 2030 target is to have 76.5% of the population living within 15 min to a public park. Houston ranks 79th in the country on this indicator.

The particulate matter (2.5) concentration in the Houston metro was averaged at 11.7 in 2015. The range in the USA was from the best performing metro Palm Bay, Florida, with a PM2.5 concentration of 6.5, to the worst performing metro Riverside, California, with a PM2.5 concentration of 14.8. The 2030 target for this indicator is to reduce PM2.5 concentrations to 6.3 $\mu g/m^3$. Houston ranks 85th in the country on this indicator.

The ozone level reported for the Houston metro was 0.077 ppm in 2016. Houston is slightly higher than the national average of 0.071 ppm. The best performing metro was Urban Honolulu, which had a reading of 0.048 ppm. The 2030 goal is to reduce ozone concentrations to 0.05 ppm, which is a standard recommended by the World Health Organization (WHO). Houston ranks 84th in the country on this indicator.

4.4.12 SDG12: Responsible Consumption and Production

SDG12		Data	N	Minimum	Maximum	Mean	Houston
	Toxic releases	lbs per square mile	100	3.5	33,728.9	1432.9	**9601.7**

Source: (Espey et al. 2018). Toxic releases 2016 data

Toxic industrial waste released into the air, water, or land was used as the metric for this indicator. The Houston metro reported 9601.7 lbs per square mile of toxic releases. This was higher than the national average among metros of 1432.9 lbs/sqml. The best performing metro was McAllen, Texas which reported only 3.5lbs/sqml of toxic releases. Houston ranks 97th in the nation on this measure. The global 2030 target is to reach 13.4lbs/sqml of toxic releases.

4.4.13 SDG13: Climate Action

SDG13		Data	N	Minimum	Maximum	Mean	Houston
	Household carbon footprint	TCO2e per capita	100	12.4	51.4	22.9	**33.7**

Source: (Espey et al. 2018). Household carbon footprint 2013

The Houston metro's reported total carbon emissions per capita was 33.7 in 2013. The average among metros in the nation was below Houston at 22.9. The best performing metro was Akron, Ohio with a reported 12.4 carbon emissions equivalents per capita. Houston ranks 89th in the nation on this measure. The 2030 target is to reduce carbon emissions per capita to 1.7.

4.4.14 SDG15: Life on Land

SDG15		Data	N	Minimum	Maximum	Mean	Houston
	Open space	sq-meters per capita	87	14.9	226.0	84.5	NC
	Brownfield and Superfunds	Sites per square mile	100	0.0	0.5	0.1	0.04

Source: (Espey et al. 2018). Open space 2016 data. EPA CleanUp sites 2018 data

The Detroit, Michigan metro area leads the nation with 226 square meters of open space per person. The average among US metros with reported data was 84.5 square meters per person. The 2030 SDG target is to reach 187 square meters of open space per person. There was no reported data for the Houston metro on this measure.

The average metro area in the USA has a reported 0.1 EPA Brownfield or Superfund Site per square mile. Three metro areas have no reported EPA Brownfields or Superfund Sites; those are Provo, Utah; Bakersfield, California; and Colorado Springs, Colorado. The Houston metro area has a reported 0.04 Brownfield or Superfund Site per square mile. Houston ranks 49th in the nation on this measure. The 2030 target is to reduce this number to 0.004 sites per square mile.

4.4.15 SDG16: Peace, Justice, and Strong Institutions

SDG16		Data	N	Minimum	Maximum	Mean	Houston
	Violent crimes	1/100000	92	8.5	1262.3	425.3	**578.2**
	Firearms deaths	1/100000	100	2.1	25.6	10.3	**11.6**

Source: (Espey et al. 2018). Violent crimes 2016 data. Firearm deaths 2014 data

The average number of violent crimes per 100,000 people in the USA was 425.3 in 2016. The Houston metro reported 578.2 violent crimes, which was above the national average. The best performing metro was Youngstown, Ohio which reported only 8.5 violent crimes for every 100,000 persons. Houston ranks 75th in the nation on this measure. The 2030 target is to reduce the number of violent crimes per 100,000 persons to 71.3.

The average number of deaths by firearms per 100,000 people was reported at 10.3 in 2014. The Houston metro area reported only slightly higher than this number with 11.6. The best performing metro area in the country was Urban Honolulu with 2 deaths per every 100,000 persons. On the mainland, the Boston, Massachusetts metro reported the next lowest number of 3.56 deaths per every 100,000 people caused by firearms. Houston ranks 66th in the nation on this metric. The 2030 SDG target is set at 3.5 deaths for every 100,000 people.

4.5 Conclusion

To develop solutions for living and managing a sprawl region, we should actually revisit the quandary, which propelled George Mitchell to convene sustainability leaders in Houston back in 1974. The quandary was the question of growth vs development. Herman Daly suggested that an economy based on endless growth in physical production was impossible, because the world is finite, and therefore a steady state is a physical necessity. To attain the proposed state of economy, Daly suggested his concept of a steady-state economy based on qualitative but not quantitative growth, taking into consideration four important questions: (1) At what levels should the stocks of wealth and people be maintained constant? (2) What is the optimal level of maintenance throughput for a given level of stocks? (3) What is the optimal time horizon or accounting period which population and wealth are required to be constant? (4) What is the optimal rate of transition from the growing economy to the steady state? To achieve "optimum rates," he proposed a "zero growth" scenario and also suggested to control the right to have children in order to address the population problem and the control of the consumption of different resources. He recommended a system where the government would basically auction off the right to consume basic resources, and the time aloud to be consumed would decrease over time, having in place market mechanisms such as higher prices, conservation, better tech, substitutions, etc. as a way to limit the consumption (Daly 1973). Mitchell was inspired by Daly's ideas but was inspired to host the conferences to engage leaders in discussing and developing alternative solutions since he was philosophically opposed to the "No-Growth" scenario, which Daly presented. For Mitchell it was more of a question of how to grow, rather than limiting growth (Schmandt 2010).

Mitchell spent a great deal of money and resources to address this issue of growth vs development. Based on the performance of the Houston region, the dilemma is still very acute and its adverse effects are growing larger. Our research team will engage regional business leaders to further Daly's concerns complemented by Mitchell's interest to develop solutions and interventions to address the question, "How can we grow sustainably in a sprawling region?"

We end this chapter with a few takeaway points from Houston's performance in comparison to the 100 largest metros in the country:

- SDG1 No Poverty, SDG4 Quality Education, SDG 8 Decent Work and Economic Growth – With a poverty rate above the national mean, we have to address the determinants of low economic contribution and poverty. Education is a major determinant to these areas. Professional organizations such as American Leadership Forum are organizing members to develop interventions for the wicked problem of education in Houston. As the state takes over the school district starting in 2020, the business community should support efforts by intervening to ensure student outcomes support industrial needs in the region.
- SDG3 Good Health and Well-Being – Heart attacks, diabetes,[3] traffic accidents, and access to physicians can also be correlated with effects of living in a sprawl city. These areas will all need public intervention to improve our regional performance. Lack of health insurance is both a national- and state-level political issue, which may correlate strongly with health issues such as infant mortality, low birth weight, and STDs.
- SDG6 Water Stress – The City of Houston is a regional water provider, which is good for the residents within the City of Houston; however it can become burdensome on city administration unless appropriate monies are charged to support the operations of this resource.
- SDG7 Affordable and Clean Energy – This is another area in which business needs support from local government to incentivize clean energy sources. Incentives may be in the form of loans or grants for solar panel installation or loans to incentivize business to develop more clean energy sources.
- SDG10 Reduced Inequalities; SDG16 Peace, Justice, and Strong Institutions – As our region grows in terms of the Hispanic ethnicity, we must develop interventions to ensure wicked inequalities do not reduce our economic success or our quality of life in general. We need to do more to ensure that all historic minorities have adequate opportunities based on their needs to be able to develop without prejudice.
- SDG11 Sustainable Cities and Communities, SDG12 Responsible Consumption and Production, SDG13 Climate Action, SDG15 Life on Land – As a result of its large land area, the Houston region has to develop policies and interventions to deal with the effects

[3]Diabetes may develop from eating the wrong types of foods since access to fresh fruits and vegetables are limited in a sprawl city.

of sprawl. These issues include transportation efficiency, air pollution, and health and well-being impacts to commuters. Houston area governments and business stakeholders have to address fragmented governance for subregions which share common industry such as the Port of Houston and the Texas Medical Center. Fragmented governance also needs to be addressed not only for geographic subregions but for regional issues such as flood management and food access. In the flood management area, there needs to be more formalized collaboration between the City of Houston, Harris County, and Montgomery County. These are the three major jurisdictions within the San Jacinto Watershed, which is the geographic determinant of flood hazard in the region. As in the case of the Port of Houston, one solution may be for the state to create a regional body of commissioners and conduct periodic reviews of the governing body. In the case of food access, governance in the region needs to adopt a more formalized role to ensure accessibility. Public-private partnerships to support locating supermarkets in places, which market determinants alone cannot support, are absolutely necessary to improve food access. One major effect from sprawl is that businesses which depend on density thresholds, such as supermarkets, will always need support to locate across the sprawl city.

References

Abraham D, Race B, De Leon A (2019) Mid-Brays resiliency & revitalization housing & community development strategies. Rice University, Houston. http://hdl.handle.net/1911/107432

Blackburn J (2011) Measuring city sustainability: project Houston. Rice University, Houston

Daly H (1973) Towards a Steady-State Economy. W.H. Freeman, New York

EPA (2001) Our built and natural environments: a technical review of the interactions between land use, transportation, and environmental quality. Environmental Protection Agency, Washington, DC

Espey J, Dahmm H, Manderino L (2018) Leaving No U.S. city behind: The 2018 U.S. cities SDGs index. Sustainable Development Solutions Network, New York

Florida R (2002) The rise of the creative class. Basic Books, New York, NY

GHP (2019) The economy at a glance

HGAC (2011) Our great region 2040, Houston

King L (2012) Houston sustainable development indicators: a comprehensive development review for citizens, analysts and decision makers. Shell Center for Sustainability, Rice University, Houston. https://hdl.handle.net/1911/71239

King L (2013) The sustainable development of Houston districts: the health of the city report. Rice University, Houston. https://hdl.handle.net/1911/75737

King L (2014) Houston community sustainability: the quality of life Atlas. Rice University, Houston. https://hdl.handle.net/1911/107590

King L (2015) The 5th Houston sustainability indicators report: opportunity among the Houston districts. Rice University, Houston. https://hdl.handle.net/1911/107588

King L (2016) The 6th Houston sustainability indicators report: sustainable communities and public education. Rice University, Houston. https://hdl.handle.net/1911/107589

King, Lester. 2017. "Comprehensive sustainability indicators: The Houston sustainability indicators program." In *Community quality of life indicators: best cases VII*, by Meg Holden, Chantal Stevens and Rhonda Phillips, 167–183. Cham: Springer

Kotkin J (2007) Opportunity urbanism: an emerging paradigm for the 21st century. Greater Houston Partnership, Houston

Pendall R, Puentes R, Martin J (2006) From tradition to reformed: a review of the land use regulations in the 50 largest metropolitan areas. Brookings Institute, Washington, DC

Sanborn B (2012) The region's high schools. 2012 Community Indicators Symposium. Center for Houston's Future, Greater Houston Partnership, Houston

Schmandt J (2010) George P Mitchell and the idea of Sustainability. Texas A&M University Press, College Station

SCS (2017) Shell center for sustainability. Rice University. https://scholarship.rice.edu/handle/1911/71237. Accessed 17 Sept 2019

SGN (2002) Getting to smart growth: 100 policies for implementation. Smart Growth Network, Washington, DC

Sustainability Solutions Lab (2018) Houston sustainability indicators. https://www.HoustonCommunitySustainability.org/. Accessed 6 Nov 2019

TRB (1995) Expanding metropolitan highways: implications for air quality and energy use. Transportation Research Board, Washington, DC

Wallace J (2019) Money grab. Mayor Turner and Gov. Abbott clash over Harvey recovery. Houston Chronicle, October 9

Wooster J (2013) The business advantage of sustainability. Plast Technol 59(3):36–41

David B. Abraham, PhD (co-editor) is a scholar of Environmental Planning and Policy, specializing in Sustainability Planning and Performance Measurement. His research interests focus on developing sustainability strategies for the enhancement of living places. His research and professional experience includes food insecurity planning, health and well-being impacts from the built environment, urban forestry sustainability strategies, resiliency and revitalization community planning, and community performance metrics. His work has been featured at workshops, keynote agendas, and conference proceedings throughout the USA and internationally in Ireland, Barbados, Brazil, Canada, and Portugal. Dr. Abraham is a research scientist and lecturer at Rice University.

5 Making the SDGs Relevant for Cities: Using the Community Capital Tool in British Columbia

Maria Spiliotopoulou and Mark Roseland

5.1 Introduction

The scientific evidence on the Earth's deteriorating condition – and the urgency to respond with effective action – has been mounting for decades. The increased frequency of extreme phenomena; the persistent poverty, increasing social and economic inequality, and inaccessibility to basic provisions; the decline of ecosystem services; and the unprecedented species extinction are some of the signs that the Earth may soon not be able to sustain the growth of human population and economic activity while maintaining systemic planetary well-being (Daly 2005; Steffen et al. 2011).

From the 1987 UN World Commission on Environment and Development report "Our Common Future" (Brundtland) to the 1992 UN Conference on Environment and Development in Rio, then the 2002 Johannesburg World Summit on Sustainable Development, followed by the 2012 Rio+20 Earth Summit, then the 2015 Sustainable Development Goals (SDGs), the 2015 Paris Climate Accord, and most recently the 2017 New Urban Agenda, the message has been loud and clear: the world needs to be on a more sustainable pathway, quickly, if we are to have any hope of a sustainable future. Yet effective action, as well as political will, has been elusive. One reason for this is because these global challenges must be addressed at national and local levels.

In this chapter, we present our case studies with two municipalities in British Columbia, Canada, where we applied modified versions of the Community Capital Tool (or CCT, detailed below) and conducted a complex matching and mapping exercise to show the relationship between the SDGs, the CCT, and local goals in the municipalities. We also discuss the challenges and opportunities we identified with regard to achieving local sustainability goals and contributing to Canada's commitments toward the UN Global Goals.

5.1.1 Sustainable Development Globally

The 1987 Brundtland Commission report noted the interconnectedness between human activity and environmental degradation: 26% of the world's population, living in developed countries, consumed 80–86% of nonrenewable resources and 34–53% of food products (WCED 1987). The increased frequency of extreme phenomena and the persistent social and ecological issues such as poverty and decline of ecosystem services have led a growing number of scholars to

M. Spiliotopoulou (✉)
Simon Fraser University, Vancouver, BC, Canada
e-mail: mariaspi@sfu.ca

M. Roseland (✉)
Arizona State University, Tempe, AZ, USA
e-mail: Mark.Roseland@asu.edu

D. B. Abraham, S. D. Iyer (eds.), *Promoting the Sustainable Development Goals in North American Cities*, Sustainable Development Goals Series, https://doi.org/10.1007/978-3-030-59173-1_5

refer to the modern period as "the Anthropocene." This is defined as the era marked by the detrimental impact of human activity on the planet (Steffen et al. 2011). We no longer live in an "empty world" (empty of us and our waste), but rather in a "full" one (Daly 2005), with significant implications and repercussions for current and future generations.

The necessity for limits to growth, initially expressed in the 1970s, is now strongly supported by up-to-date research on planetary boundaries that have been exceeded, such as genetic diversity and climate change (Steffen et al. 2015; Hamstead and Quinn 2005; Meadows et al. 1992). Current generations now have both the knowledge and the responsibility to lead humanity away from putting further pressure on the planet and toward a safer and more sustainable future (Rockström 2009; Steffen et al. 2011).

In this spirit, in 2000, the UN Member States adopted the *Millennium Declaration* aspiring to eradicate extreme poverty and reduce inequalities, with a particular focus on developing countries; the Global North would mostly contribute to development aid and financing. The Millennium Development Goals (MDGs) were composed of 8 goals, 21 targets, and 60 indicators and encouraged action by a broad range of stakeholders in an effort to address the multidimensional issue of extreme poverty by 2015.

Several of the goals were achieved in the 15-year period in the developing world, with notable decreases in extreme poverty, child and maternal mortality, and disease rates and rising rates of primary school enrollment and of life expectancy (United Nations 2015c). Severe problems however persisted in areas such as sub-Saharan Africa and South Asia, because of the extensive slums and limited access to freshwater, sanitation, and medicines. The MDGs were generally criticized as vague, disconnected from a whole-system view, difficult to measure (partly due to data insufficiency), and potentially causing further inequality in urban areas (Harcourt 2005; Meth 2013).

Building partly on the achievements of the MDGs but mainly acknowledging the continuing struggles in social, economic, and ecological systems around the world, the Sustainable Development Goals (SDGs) were unanimously and ceremoniously approved by the UN Member States in September 2015 (United Nations 2015b). The UN 2030 Agenda for Sustainable Development, which includes the 17 Global Goals (SDGs) and 169 concrete targets, is a significant step forward and a turning point for global sustainability.

Despite the long consultation and negotiation process (more than 3 years), the initial promoters of an inclusive agenda (Colombia, Guatemala, Peru, and United Arab Emirates) achieved their objective: that the SDGs address *sustainable* development and not simply development (often confused with growth) like their predecessors, the MDGs (Dodds et al. 2017). The new goals offer a more integrated vision and plan for this millennium: they apply to both developed and developing nations; and they are grounded in a holistic, systemic view of sustainability (Woodbridge 2015). The acknowledgment that the principal global challenges, this century (ecological integrity, social equity and cohesion, and economic prosperity), need to be addressed in a holistic way is also reflected in the 2015 UN Climate Change agreement (United Nations 2015a) and in the UN New Urban Agenda (United Nations 2017).

Achieving the 17 SDGs with their 169 targets and numerous associated indicators is a complex undertaking that must be addressed at numerous scales from global to local. In response, we have framed our research to focus on the full set of SDGs at the local scale as a way to address, monitor, and achieve the SDGs nationally and globally.

5.2 Developing and Monitoring Sustainable Communities

Since the negotiations stage, people and organizations involved in the SDGs development process stressed the importance of localizing the global goals (Dodds et al. 2017). The success of the SDGs is conditional on creating and implementing successful, monitorable, and transferable sustainability policies and practices in communities. We posit that a predominantly bottom-up or "community-up" approach is crucial for the SDGs to gain wide traction, engage citizens and other stakeholders, and ultimately succeed in turning sustainability into the new modus operandi globally, within this century.

5.2.1 Developing Sustainable Communities

Our research is situated in the field of sustainable development in local communities, with particular focus on urban areas, which are projected to be home to at least 68% of the world's population by 2050 (UN DESA 2018). For our research purposes, a community refers to "a group of people bound by geography and with a shared destiny, such as a municipality or a town," and is considered as a complex, adaptive, and interconnected system requiring interdisciplinary study (Uphoff 2014; Roseland 2012). An urban area is "a human settlement characterized – ecologically, economically, politically and culturally – by a significant infrastructural base; a high density of population, whether it be as denizens, working people, or transitory visitors; and what is perceived to be a large proportion of constructed surface area relative to the rest of the region" (James 2015).

Cities occupy 3–4% of the world's land surface, use ~80% of resources, and discharge most global waste while being increasingly vulnerable to climate change and health challenges linked to high economic and environmental costs (Kanuri et al. 2016; (Girardet 2015). The latest global urbanization projections for 2050 and the accumulation of challenges in cities prove the urgency of developing local solutions to global (or "glocal") issues. Cities have enormous productivity potential in terms not only of economic and labor productivity (diverse and inclusive economy, fostering innovation) but also of social productivity (hubs of research, learning, and sharing) and ecological productivity (ecological function regeneration and efficient use of resources) (Roseland and Spiliotopoulou 2017).

The full set of the SDGs is relevant to local communities even though the UN Global Agenda for 2030 includes a goal specifically for cities: goal 11 for inclusive, safe, resilient, and sustainable cities and human settlements (Kanie et al. 2014). Achieving long-term sustainability locally requires a focus on all goals, not just goal 11, in order for societal change through collaborative decision-making and community engagement to occur, as the principles of sustainable community development (SCD) so urge (Clarke 2012; Hermans et al. 2011). SCD is a holistic approach that integrates social, environmental, and economic considerations into the dynamic processes toward community sustainability, while providing for current and future generations (Berke and Conroy 2000; Roseland 2012).

SD and SCD have been influenced by a number of theories and have matured over the last few decades in academic, professional, and popular discourse. While SCD may be a fairly new paradigm for local development, it is rooted in such intellectual traditions of the previous two centuries as social ecology, bioregionalism, native worldviews, ecological modernization, self-reliance, eco-localism, environmental justice, etc. (Roseland 2000; Roseland and Spiliotopoulou 2016). More recently, SCD has embraced strong sustainability principles which acknowledge the Earth's regenerative limits and the need for socio-ecological and economic resilience "across temporal and spatial scales" (Meerow et al. 2016; Daly 2005; Rockström et al. 2009).

Under the strong sustainability model, social and ecological considerations are increasingly being included in community analysis and policy-making through collaborative and systemic processes. Several parallels can thus be drawn between this comprehensive paradigm for local development and the UN 2030 Agenda for Sustainable Development. These include the long-term and whole-systems perspective, the recognition of the dynamic nature of socio-ecological systems, and the potential to reveal opportunities for synergies and indirect positive impact among the various dimensions and goals for sustainability.

5.2.2 Monitoring Sustainable Communities

In pursuit of the balanced and integrated approach that SCD and the SDGs advocate, communities are challenged by the difficulties of addressing multiple objectives and monitoring their progress while setting priorities at a higher level of decision-making. They face the complexity of sustainability goal setting and the challenge of navigating the variety of local agendas grounded in diverse theoretical backgrounds or stakeholder interests (Roseland and Spiliotopoulou 2017). They also need to meaningfully engage citizens in a broad range of decision-making processes and collect data efficiently and consistently to allow for effective progress monitoring and assessment (Caprotti et al. 2017; Moreno Pires et al. 2017).

One way to address these challenges is by adopting sustainability planning and assessment frameworks and tools that inform and mobilize citizens and their governments. The assessment of plans through sustainability frameworks is considered an effective tool that follows implementation in order to gauge success and measure performance in ecological, social, and economic terms (Roseland 2012). Despite the abundance of tools and agendas, not all of them promote a whole-systems approach or assist in concrete implementation and effective monitoring (Joss et al. 2015). Successful SCD monitoring and assessment entails tackling issues such as stakeholder engagement, place-specific context, political credibility, and adoption of a shared and practical vision.

The research foundation of the pilot projects we present here is the Community Capital Framework (CCF). Its purpose is to support decision-making not only as a planning toolkit but also as a performance and progress assessment instrument. The CCF has been designed with a systems thinking perspective: each form of community capital is a subsystem of the larger whole community system. Since the early 2000s, we have used the CCF in various communities (big, small, rural, urban, developed, developing) around the world with success (e.g., in North America, Latin America, and Eastern Europe).

Built upon the CCF, the Community Capital Tool (CCT) is an SCD monitoring and assessment tool and the product of collaboration between the Centre for Sustainable Development at Simon Fraser University in Canada and Telos, the Brabant Center for Sustainable Development, Tilburg University, Netherlands (Roseland 2012). In this context, the term "community capital" includes natural, physical, economic, human, social, and cultural forms of capital (see also Fig. 5.1) (Roseland 2012):

1. *Natural Capital:* Living within ecological limits, conserving and enhancing natural resources, using them sustainably, using cleaner production methods, and minimizing waste
2. *Physical Capital:* Community assets such as public facilities, water and sanitation provision, efficient transport, diverse housing, adequate infrastructure, and telecommunications
3. *Economic Capital:* Circulating money within a community, producing locally, trading fairly, and developing community financial institutions
4. *Human Capital:* Focus on health (including food, shelter, and safety), education, family and community cohesion, and enhanced training and improved workplace dynamics
5. *Social Capital:* Effective and representative local governance, participatory planning, access to information, capacity building, safety, and collaboration and partnerships
6. *Cultural Capital:* Attention to traditions and values, heritage and place, the arts, diversity, and local history

The Tool's six capitals are broken down into a set of smaller stocks (or categories) used to measure capital capacity and sustainability progress. These stocks are universal and were chosen based on their ability to accurately represent the health of each capital. Within each stock is a set of requirements that are adaptable to the local context, needs, and priorities of the community or the context of the specific initiative being measured. Lastly, the status of each requirement is "indicated" by one or more specific, measurable indicators. The CCT then shows the final results as graphics that report on the health of

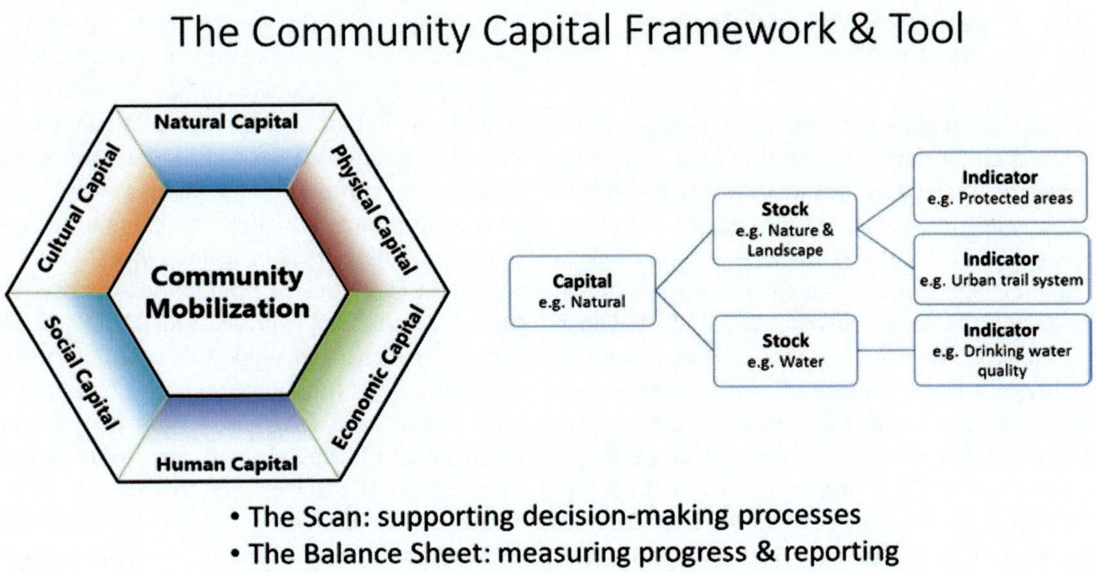

Fig. 5.1 Community capital: a framework and tool for sustainable community development. (Adapted from: Roseland 2012)

each capital account and each of their constituent stocks.

Community leaders, planners, and citizens can use this information to compare the current sustainability status of their community with past results and potentially with other comparable communities. The CCT was designed based on strong sustainability principles that advocate for the preservation of adequate amounts of all natural assets while avoiding terminal damage to critical natural assets and consciously seeking to address key social issues. It focuses on community-specific issues in a way that recognizes each community's regional and global impact on the environment and on society at large. The CCT is intended to incorporate the democratic input of citizens in terms of values and priorities and provides planners and decision-makers with a tool that helps them ensure that these values and priorities are reflected in policy decisions (Roseland 2012).

In the case studies presented here, we also consulted several other sustainability assessment frameworks. These frameworks contributed to our improved understanding of this field and played an important role in shaping the CCT for the two municipalities we worked with (see more details in the next section). These sustainability frameworks are (in no particular order) the UN Sustainable Development Goals, STAR Communities (recently merged with LEED for Cities), One Planet Living (or Eco Communities), ISO 37120, Community Foundations of Canada Vital Signs, Green City Index, Living Community Challenge, the EU Reference Framework for Sustainable Cities, LEED-ND and LEED for Cities, EcoDistricts, the International Eco-City Framework and Standards, and the City Resilience Index.

With regard to the SDGs in particular, we were able to demonstrate through our case studies that the CCT is very much aligned with the SDG framework. As we will explain in detail below, the CCT is structured in a similar way to the SDGs – they both have three levels of forward-looking decision-making (goals, targets, and indicators) – and their indicators overlap by more than 50%.

5.3 Research Methodology and Context

In this research, we engaged a mixed-methods, information-oriented approach within case study research, integrating quantitative and qualitative data collection and analysis techniques and tools (van Kerkhoff 2014; Yin 2014). For reasons of funding[1] and focus, we decided to work with two communities in the Lower Mainland of British Columbia: the City of Maple Ridge and the District of North Vancouver. Whereas some communities may see the SDGs as either irrelevant to or in conflict with local priorities, we partnered with two cities that approached us[2] and demonstrated interest in participating in our research in order to enhance their sustainability planning and performance assessment processes, while exploring common ground with the SDGs.

The main objective of this research was to help the two municipalities (Fig. 5.2) achieve their stated visions by providing them with a sustainability assessment framework that would be relevant to their needs and values, while connecting them to a broader context. The customized integrated framework would support city council, staff, citizens, and other community stakeholders in effectively identifying community needs, allocating funds, implementing policies and programs, and measuring results, from a long-term perspective.

5.3.1 Case Studies Context: The District of North Vancouver (DNV)

As one of three municipalities on the North Shore of Metro Vancouver, the District of North Vancouver (DNV) shares key infrastructure (roads and utilities) and in some cases partners in the delivery of services (recreation and emergency services). Its natural assets define the local lifestyle and values, and the industrial waterfront, a strategic national asset, provides significant business opportunities and local jobs. A growing community with two First Nations reserves, the District considers collaborative planning essential to the achievement of its long-term goals.

The DNV Official Community Plan (OCP),[3] titled "Identity 2030," presents the DNV's vision for an "inclusive and supportive community that celebrates its rich heritage and lives in harmony with nature" and that has a "network of well designed and livable centres" and "resilient and diverse" local businesses (District of North Vancouver 2011). Our project with the District was carried out in 2018 and aimed to help achieve this vision by adding to the monitoring and reporting work of the Community Planning Department and the Official Community Plan Implementation Monitoring Committee 2017–2018. Our other objective, inspired by how cities like San Jose and Baltimore localized the SDGs, was to compare the District's goals and indicators to the SDGs and their targets and indicators and to make recommendations on how to address gaps identified.

5.3.2 Case Studies Context: The City of Maple Ridge (CMR)

Located 45 kilometers east of Vancouver, Maple Ridge is a family-oriented community and one of the fastest growing cities in Metro Vancouver. It has a vibrant local economy and the most affordable industrial land and real estate in the region. It is committed to becoming a sustainable community by considering the environmental, social, and economic impacts of its actions for present and future generations. The City of Maple Ridge

[1]Please refer to the end of this chapter for a disclosure statement regarding the funding for this project.

[2]We were approached by and collaborated with the Community Planning Department in the District of North Vancouver and the Sustainability and Corporate Planning Department in the City of Maple Ridge.

[3]Under British Columbia's Local Government Act, municipalities and regional districts are encouraged to develop an Official Community Plan (OCP) that provides a long-term vision for the community. An OCP is "a statement of objectives and policies that guide decisions on municipal and regional district planning and land use management" (Province of British Columbia, n.d.).

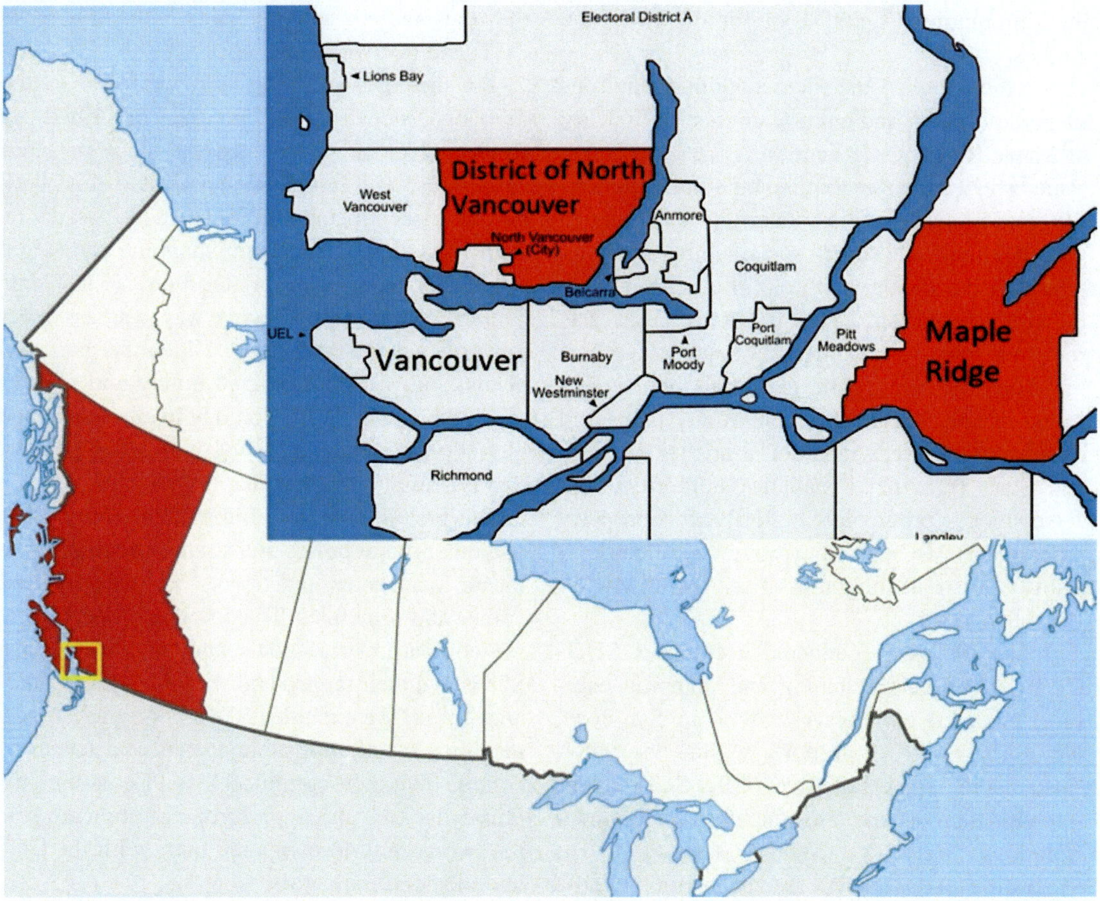

Fig. 5.2 The province of British Columbia and, in the inset, our two case study municipalities. (Images by TUBS/CC BY-SA 2.5 and by TastyCakes/CC BY 3.0)

(CMR) Official Community Plan lays out the city's long-term vision for a "vibrant and prosperous [community, with] a strong local economy, stable and special neighbourhoods, thoughtful development, a diversity of agriculture, and respect for the built and natural environment" (City of Maple Ridge 2014).

As with our other case study, the main objective of the Maple Ridge project carried out in 2017 was to help the City achieve this vision by assessing current sustainability and providing the City and its citizens with a customized sustainability assessment framework. Although the City of Maple Ridge did not at the time explicitly express interest in aligning their goals with the SDGs or taking advantage of the SDG framework in a specific way, we nevertheless used the SDG framework in the project reported in this chapter.

5.3.3 Research Methodology

Within a mixed-methods approach, we started working on the case studies by examining the related literature and particularly exploring the current arena of sustainability frameworks, tools, and best practices. We reviewed a significant number of sustainability monitoring and assessment frameworks worldwide as well as initiatives and best practices for planning and assessment in other communities in Canada and beyond. As mentioned above, the SDGs and other frameworks and tools helped inform the adjustment of the Community Capital Tool for the two case studies.

We then studied the socioeconomic, environmental, political, and cultural context in the City of Maple Ridge (CMR) and the District of North Vancouver (DNV) and collected some quantitative data to evaluate the capacity of each city to source reliable and timely sustainability data and to establish an initial picture of their current sustainability situation. This quantitative data was retrieved from various archival sources and was measured against specific set goals and targets found in policy and other community plans and documents. Sources included Statistics Canada, BC Stats (provincial statistics authority), BC Assessment (provincial authority for property assessment), BC Hydro (provincial electricity utility), local health authorities, and CMR or DNV databases.

In parallel, we conducted a complex SDG-CCT-Local Goals matching and mapping exercise, modeled on the work done in San Jose, New York, and Baltimore within the USA Sustainable Cities Initiative (USA-SCI), under the guidance of the Sustainable Development Solutions Network (SDSN) (Nixon 2016; Prakash et al. 2017). As shown in Fig. 5.3, the mapping extended along three levels of decision-making within three frameworks: we compared

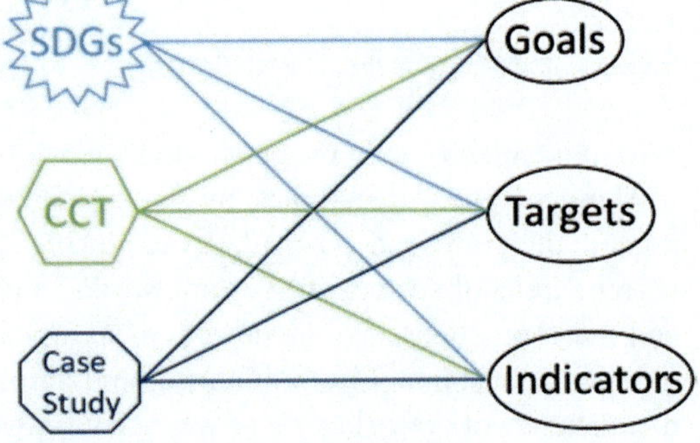

Fig. 5.3 The extensive mapping of the two cities' goals, targets, and indicators with the SDGs and the CCT. The shape of the SDG and the CCT frameworks reflects the number of goals (or capitals) included in each, excluding SDG 17 on global partnerships

the goals, targets, and indicators of the SDGs, the CCT, and the two case studies.[4]

For the SDG mapping task, we followed the first two steps of the process described by Ruckstuhl et al. (2018) and the steps in Mesa et al. (2019), although we conducted this work before these documents were made available. Step 1 was policy and target mapping and step 2 was identifying appropriate metrics and data sources. Our overall objective was to assess existing policy goals and targets, identify gaps and needs, and offer customized policy and metrics recommendations that would help align local and global goals. It is this part of the research project that is presented in this chapter in detail.

We first studied the official community plans and other major policy and strategy documents to locate local goals and targets and identify core values and principles. To complete this first step, we compared local goals and targets with the SDGs and their targets and with the CCT capitals and stocks. We excluded SDG 17 on global partnerships as not applicable at the local level and context. We then compiled lists of existing sustainability and other performance indicators in the two cities and compared them with the CCT and the SDG indicators.

In addition, we collected qualitative data through interviews, meetings, and workshops with key stakeholders in both municipalities. We engaged elected officials (councillors), appointed officials (city senior management and expert staff), and community members through the North Shore Community Foundation and the Maple Ridge Community Foundation.[5]

A series of meetings with key staff provided us with valuable perspectives on various aspects of localizing sustainability indicators; we met with departments such as Community Planning, Parks, Public Works, Economic Development, Information Technology, Engineering, and Emergency Services (Fire and Police). Through

[4]An SFU Master of Resource Management Planning student, Daniel Ross, was also involved in this part of the DNV project (Ross 2018).

[5]Community foundations manage private endowments to provide local projects with funding for initiatives that benefit the community.

these meetings, the subject-matter experts largely contributed to our understanding of indicator contextual meaningfulness, policy jurisdiction, data availability, data sources, existing targets, municipal capacity, etc.

In the DNV, we also engaged with the 2017–2018 Official Community Plan Implementation Monitoring Committee (OCP IMC) which is composed of community members and whose purpose is to provide comments on OCP implementation (consistency of vision, goals, and actions), monitoring (ensuring meaningful and appropriate indicators), and communication with the public.

In total, we conducted 14 interviews in the DNV and 16 in the City of Maple Ridge, we consulted more than 20 subject-matter expert staff in each municipality, and we engaged more than 40 community members in workshops with the 2 community foundations and the OCP IMC. Thanks to this inclusive participatory process, we had the opportunity to explore and identify perceptions of community stakeholders on needs and gaps and document their preferences and ideas regarding the linkages between global and local sustainable development, assessment tools, and visions for the future; we also received their direct feedback for our work on sustainability frameworks. Figure 5.4 illustrates the methodological model of the participatory process used in both case studies.

5.4 Research Findings

In our case studies in the Greater Vancouver area, we applied sustainability assessment methods and tools to support the DNV and the City of Maple Ridge increase their sustainability potential and identify synergies with the SDGs. This section presents, firstly, the research findings from the mapping of local goals, targets, and indicators with the SDG framework and the Community Capital Framework and Tool and, secondly, our findings from analyzing the interviews with focus on connections between global and local agendas.

5.4.1 Goals, Targets, and Indicators Mapping

Following the examples of New York, San Jose, and Baltimore, we performed a complex mapping and alignment exercise in the DNV and in the CMR (Nixon 2016; Prakash et al. 2017). We compared the higher-level goals of both cities with the SDGs, their targets (or "stocks" in the case of the CCT) with the SDG targets, and their – at the time current – indicators with the SDG and the CCT indicators.

With regard to the DNV, through this work we observed that the eight major goals or objectives in the DNV's Official Community Plan were aligned with seven SDGs fully or quite extensively, as well as with all six capitals of the CCT. Emphasis in the DNV is mostly placed on issues of economic growth and well-being, protection of the natural environment, affordability, food security, and education infrastructure investment. As shown in Table 5.1, SDGs 3, 8, 9, 10, 11, 14, and 15 were fully covered by the DNV's goals; SDGs 1, 2, 4, 6, and 13 were partly covered; and for some SDGs (5, 7, 12, and 16), there was no explicit mention in the DNV OCP goals. SDG 17 was considered not applicable. Given the OCP's objective to guide the DNV toward a "sustainable future" by 2030, the wide alignment between local and global goals seems to indicate that sustainability principles and aspirations are important to the DNV and its citizens.

In CMR, our findings were somewhat similar to those in DNV. The Maple Ridge OCP includes a long-term vision statement and 45 principles that were approved following extensive citizen and stakeholder consultation. The OCP and other major policy documents mostly emphasize SDGs 2, 8, 11, and 15, while being partly aligned with SDGs 3, 4, 6, 13, and 16 (Table 5.1). The CMR mapping analysis demonstrated a specific focus on food security, education infrastructure investment, and making the city resilient in preparation for climate change impacts. Unlike the DNV goal alignment though, it is clear that the higher-level goals in Maple Ridge are not aligned with SDGs that promote innovation and industrial – or generally economic – infrastructure and action for

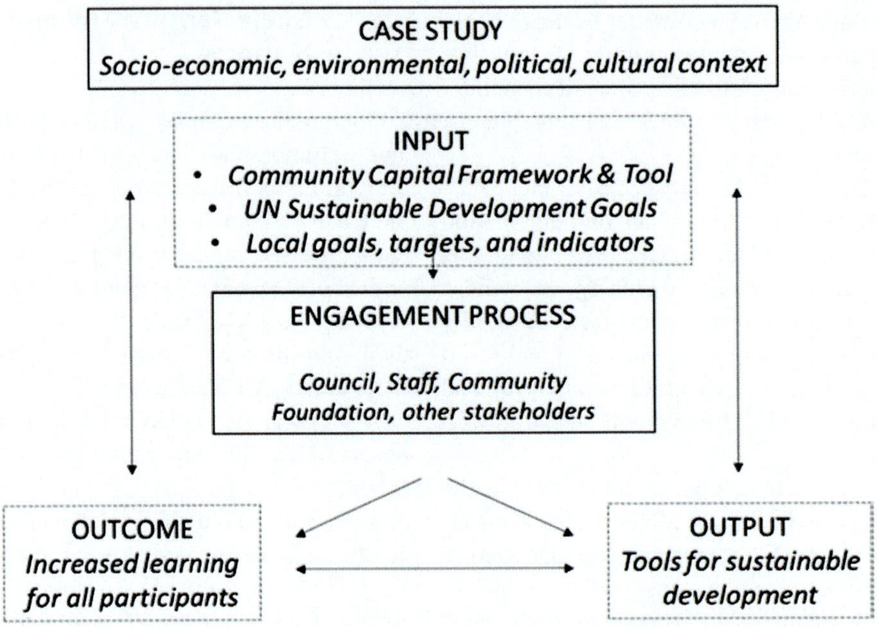

Fig. 5.4 Contextual and methodological model of participatory process. (Adapted from Hermans et al. 2011)

inequalities reduction. It makes sense however that the CMR has not placed importance on goals or targets related to SDG 14 (Life below water), since Maple Ridge is not by the ocean and therefore ocean and marine life protection are not within the city's priorities.

At the target level, we found out that although the DNV's higher-level policy documents contained a lot of recommendations and broad statements for the future, very few seemed to correspond to actionable, measurable targets. We identified 20 targets in the DNV OCP and other major policy documents, such as the Transportation Plan, the Parks and Open Spaces Strategic Plan, the Rental and Affordable Housing Strategy, the Climate Change Adaptation Strategy, and the Integrated Stormwater Management Strategy. These 20 targets corresponded to 18 (out of 169) SDG targets that are related to SDGs 1, 2, 3, 6, 8, 9, 10, and 11. In CMR, the picture is similar: we identified ten targets in the OCP, the Parks, Recreation and Culture Plan, and the Environmental Management Strategy. These ten targets correspond to only five SDG targets which are part of SDGs 6, 11, and 12.

Despite the partial alignment at the goals level, the result from the indicators mapping was different, as shown in Fig. 5.5. DNV's 26 indicators monitor progress of OCP goal implementation and range from urban growth management and park/open spaces to economic development, transportation, and climate action (District of North Vancouver 2011). These 26 indicators covered only 11.9% of the SDG indicators. We excluded 115 SDG indicators that were deemed not applicable in the District context since the SDG framework is mainly oriented toward countries. However, even after excluding those 115 SDG indicators, DNV indicators still covered only 25% of the remaining SDG indicators that were applicable. In contrast, the CCT indicators pool overlaps with the SDG indicators by more than 53%.

The City of Maple Ridge indicators mapping, on the other hand, is consistent with DNV results. CMR measures progress and performance across 69 indicators[6] ranging from energy efficiency and transportation safety and accessibility to

[6]The City of Maple Ridge calls its indicators "scorecards."

Table 5.1 Level of alignment between the SDGs and the higher-level goals of the District of North Vancouver and those of the City of Maple Ridge

SDG	DNV Goal alignment	CMR Goal alignment
1: No Poverty	Indirect match / Partly aligned (affordability & well-being)	No match / Not aligned
2: Zero Hunger	Indirect match / Partly aligned (food security)	Direct match / Fully aligned
3: Good Health and Well-being	Direct match / Fully aligned	Indirect match / Partly aligned (social services)
4: Quality Education	Indirect match / Partly aligned (education infrastructure)	Indirect match / Partly aligned (education infrastructure)
5: Gender Equality	No match / Not aligned	No match / Not aligned
6: Clean Water and Sanitation	Indirect match / Partly aligned (stormwater management)	Indirect match / Partly aligned (sensitive area protection)
7: Affordable and Clean Energy	No match / Not aligned	No match / Not aligned
8: Decent Work and Economic Growth	Direct match / Fully aligned	Direct match / Fully aligned
9: Industry, Innovation and Infrastructure	Direct match / Fully aligned	No match / Not aligned
10: Reduced Inequality	Direct match / Fully aligned	No match / Not aligned
11: Sustainable Cities and Communities	Direct match / Fully aligned	Direct match / Fully aligned
12: Responsible Consumption and Production	No match / Not aligned	No match / Not aligned
13: Climate Action	Indirect match / Partly aligned (GHGs & renewable energy)	Indirect match / Partly aligned (various related objectives)
14: Life Below Water	Direct match / Fully aligned	No match / Not aligned
15: Life on Land	Direct match / Fully aligned	Direct match / Fully aligned
16: Peace and Justice Strong Institutions	No match / Not aligned	Indirect match / Partly aligned (inclusiveness)
17: Partnerships to achieve the Goals	Not applicable	Not applicable

Red color shows no alignment, orange shows indirect or partial alignment, and green shows direct or full alignment

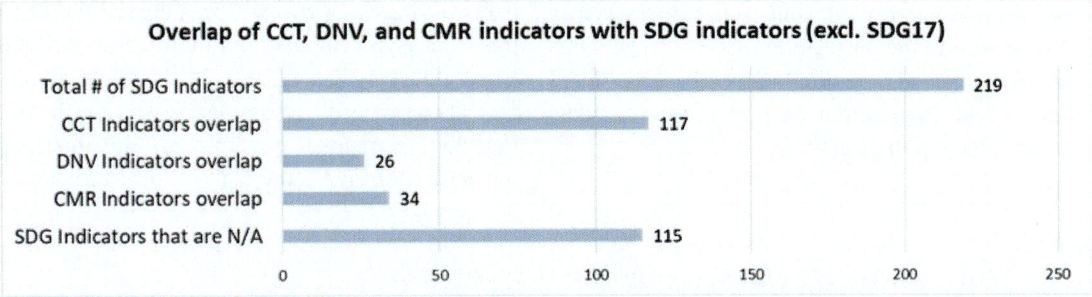

Fig. 5.5 Demonstrates the extent to which existing indicators in DNV, CMR, and CCT overlap with and address SDG indicators (excluding SDG 17 on global partnerships and showing the 115 SDGs that were considered as "not applicable")

municipal finances and emergency services efficiency (City of Maple Ridge, n.d.). There is a 15.5% overlap between CMR and SDG indicators if we take all 219 SDG indicators into consideration, but the overlap percentage increases to 32.7% if we do not include the 115 SDG indicators that we considered not relevant or applicable in the CMR context (Fig. 5.5).

5.4.2 Interview Findings

The interviews analysis through the SDG lens involved two sets of data: (1) mixed quantitative and qualitative data in response to a question on SDG awareness and familiarity and (2) entirely qualitative data in response to the open-ended question on perceptions of impact of the SDGs on local decision-making and other perceptions regarding glocal-local connections.

The majority of those interviewed were either not aware of the existence of the SDGs (50% of interviewees) or could vaguely recall having heard of them (30% of interviewees) (Fig. 5.6). Most responses were a simple "yes" or "no" but some contained additional comments. Particularly the interviewees recalling the SDGs "vaguely" or "slightly" commented that they could not cite the SDGs or that they did not have "in-depth awareness" and they were not really familiar with the details of the UN Agenda for Sustainable Development.

Responses to the second question on perceptions of SDG impact on local decision-making and on other linkages between the global and local levels yielded an extensive amount of qualitative data. A common view among interviewees was that any global goals or international commitments would probably not have a high influence on local politics and processes. While talking about this viewpoint, some interviewees attributed it to the perception that global agendas are disconnected from the local context and local beliefs and thus cannot be taken into consideration in local policy-making.

Overall, 3 broad themes – or rather problems – emerged from the analysis of the 30 interviews as far as the potential for SDG impact is concerned: the difficulty of ensuring widespread awareness and education on nonlocal matters, the issues caused by a complex public administration system involving multiple and interdependent levels of government, and the lack of accountability due to the usually nonbinding nature of international agreements.

Regarding awareness and education about the global goals and their impact, we identified a number of issues. Firstly, a few participants seemed to confuse the SDGs with other intergovernmental treaties or declarations, for instance, with the Paris climate agreement or other United Nations reports or protocols. Also, some participants who admitted not being familiar with the SDGs argued that international goals and agreements are in conflict with local goals and priorities. One interviewee mentioned that asking local governments who face serious problems (hunger, poverty, lack of clean water) to think or act globally can be perceived as distraction. That said, several participants felt that it was strange that they did not know of the SDGs although they had at some point been involved in sustainability projects or in international processes.

The second recurrent theme in the interviews was the position that the complexity of local decision-making processes is a crucial factor for the local governments not embracing the SDGs. Most interviewees talked about the challenges that accompany this complexity and the limited jurisdiction of local governments in Canada or, particularly in B.C., the lack of local jurisdiction

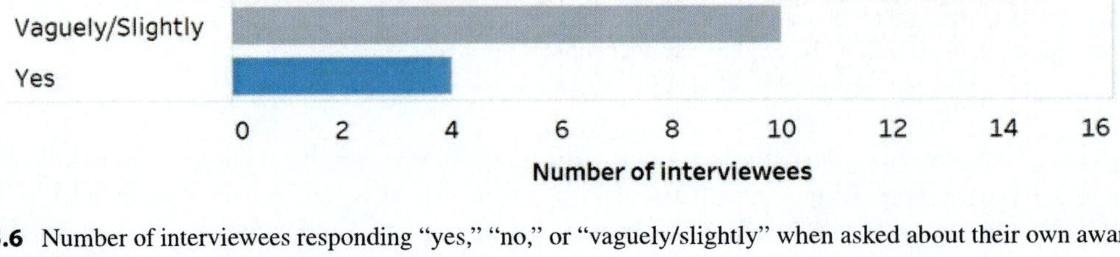

Fig. 5.6 Number of interviewees responding "yes," "no," or "vaguely/slightly" when asked about their own awareness of the UN SDGs

in some matters. These participants mainly referred to the constant struggle of local governments to secure funding from higher-level jurisdictions and the limited mandate local governments are given from provincial or federal legislation especially when it comes to important issues such as education, transportation, energy, and – to some extent – housing. Two interviewees concluded their argument on the difficulties of the multiple levels of government by describing local governments as "creatures of the province."[7]

A third SDG-related theme that surfaced throughout the interviews concerned the (usually low) level of accountability or obligation that is attached to the SDGs – or any other international agreements for that matter. The majority of interviewees expressed the belief that local governments are removed or distanced from the obligation the federal government has to achieve the SDGs and report on them by 2030. Some pointed out that local governments feel that they are more accountable to their citizens than to any national or international organizations.

At least a quarter of the participants considered the SDGs, the Paris climate agreement, and other international agreements as purely "aspirational." Some argued that setting global or local goals can be beneficial but it is inadequate when implementation is accompanied by little or no accountability. Lack of accountability in this case means that municipalities have no legal obligation, they may receive no mandate or funding to achieve the goals, and therefore they do not face any real consequences if goals are not met.

On a more positive note, it is worth highlighting that about one third of participants were explicitly in favor of embracing national or global goals and ensuring their increased impact on local decision-making. They suggested that local governments should try to align more with global goals which can provide some framework and foster connection to a wider – national and even global – context. In a few interviews, it was stated that the SDGs can present an opportunity for local governments to receive funding and other resources to achieve their local community goals. Similarly, it was argued that high-level goals such as the SDGs can help inform local decision-making and action, mainly by pointing to best practices and opportunities for learning.

Overall, our research findings indicate that there is low awareness around the UN Agenda for Sustainable Development and that recognition of the significance of the SDGs at the local level is progressing at a slow pace. The SDGs and other international agreements, despite being nonbinding, present some opportunities that are being slowly acknowledged in Canadian communities, often by municipal staff at first and then by elected officials who in general may not be that immersed in international developments. As one interviewee concisely put it, municipalities are not likely to be "driven" by the SDGs or compelled to achieve them, but perhaps may use them if they need support to achieve their local goals.

5.5 Generalizations

For this study we performed a complex SDG-Local Goals mapping exercise, similar to those undertaken in New York, San Jose, and Baltimore, and compared the goals, targets, and indicators of the SDGs, the municipalities, and our own Community Capital Tool. We also documented policy gaps and stakeholder perceptions and asked whether global agendas influence local decision-making.

Mapping the CMR and DNV OCPs and other master plans with the SDGs and the CCT was not an easy task because of the multiplicity of official documents in the two municipalities: official community plans, various sectoral plans, business plans, community sustainability plans, climate action plans, etc. Some high-level policy documents overlapped, whereas in some cases the OCP predated newer plans, and this resulted in goal and target tracking difficulties and occasional inconsistencies (Ross 2018). This mapping task however revealed significant gaps in policies and objectives in both case studies, such as low consideration of wider national or global

[7]This is indeed the case in B.C. Unlike in the USA, for example, local governments in Canada are in fact and in law "creatures of the province" they are located in.

context, fragmented prioritization in policy-making and implementation, and little attention to whole-systems integrated thinking.

The interview data offered similarly important insights, particularly into the perceptions of local elected and appointed officials about global-level goals and international commitments. What seems as a simultaneously interconnected and distanced relationship between the multiple levels of government in B.C. provides a telling argument for the lack of interest or comprehension of the UN Global Agenda. The analysis also revealed a misconception that "localizing the SDGs" absolutely requires full awareness, or even in-depth understanding, of global issues and problems in other parts of the world.

A viewpoint expressed by some interviewees is that the SDGs could be an opportunity for local governments to receive funding and other resources from higher levels of government and could offer an engaging way of approaching the potential of the SDGs. The SDGs could help initiate change at the local level even if they are perceived as an intermediate means to achieve a community's overarching goal, i.e., high quality of life and well-being for its citizens.

These findings are a strong indication of the imperative to inform and educate local governments and their citizens about the SDGs so that the latter hold the former accountable for local, national, and global commitments. Thanks to their versatile structure, the SDGs can equip communities with a broad and holistic framework for all levels of decision-making, from identifying core values, setting goals, and forming partnerships to inclusive implementation and assessment (Mesa et al. 2019).

5.6 Recommendations

In accordance with the above findings, our recommendations to both municipalities revolved around a customized comprehensive framework with a set of forward-looking and holistic-thinking indicators based on the SDGs and our research with the CCT. As mentioned above, the CCT conceptualizes communities as place-oriented, scalable, dynamic systems and is rooted in a framework that considers effects on six mutually reinforcing forms of capital: natural, physical, economic, human, social, and cultural. The Tool includes two complementary instruments: (1) the Community Capital Scan, a dialogue- and decision-support tool, and (2) the Community Capital Balance Sheet, a more rigorous quantitative assessment tool. Both are grounded in a whole-systems, integrated thinking and are structured in a very similar way to the SDGs.

The integration of the adapted version of the CCT in the two B.C. cities' decision-making processes can significantly help them achieve their sustainability goals while becoming ambassadors for SDG implementation in Canada and beyond. The Tool is a good fit to help localize the SDGs in all stages of the decision-making and monitoring process using a contextually relevant approach: firstly by expanding awareness about the global goals and increasing stakeholder participation, transparency, and perception of accountability; then by facilitating long-term goal setting and development of detailed, short-term implementation actions; and finally by supporting a locally focused but globally looking process of monitoring progress, reporting, and evaluating.

To mobilize action toward implementing and monitoring the SDGs locally, the CCT can help local officials and citizens align their goals with each other and with the full set of the SDGs while achieving effective synergies and efficiencies between goals and actions. The CCT can offer the policy roadmap and the data and visualization platform required to plan for sustainability, monitor progress, and operationalize the SDGs in the holistic and systemic spirit the SDGs themselves promote.

The B.C. municipal experience described here demonstrates that if Canadian cities incorporate tools such as the CCT into their regular practice, they can contribute to and become leaders in the achievement of Canada's Federal Sustainable Development Strategy which reflects Canada's commitment to the SDGs (Roseland and Spiliotopoulou 2018). We have every reason to expect that tools and approaches such as the CCT could work as well in other countries. Given the scale of the global sustainability challenges before us, developing these scalable and integrated local solutions may indeed provide a much-needed reason for hope.

Disclosure Statement The authors are grateful that part of the work that contributed to this chapter was funded by Mitacs Accelerate[8] (project numbers IT09978 and IT09305).

References

Berke PR, Conroy MM (2000) Are we planning for sustainable development? J Am Plan Assoc 66(December):21–33. https://doi.org/10.1080/01944360008976081

Caprotti F, Cowley R, Datta A, Broto VC, Gao E, Georgeson L et al (2017) The new urban agenda: key opportunities and challenges for policy and practice. Urban Res Pract 10(3):367–378. https://doi.org/10.1080/17535069.2016.1275618

City of Maple Ridge (2014) Maple Ridge Official Community Plan. Retrieved from https://www.mapleridge.ca/316/Official-Community-Plan

City of Maple Ridge (n.d.) City of Maple Ridge Scorecards. Retrieved January 16, 2019, from https://www.mapleridge.ca/787/Scorecards

Clarke A (2012) Green municipal fund passing go: moving beyond the plan. Ottawa

Daly HE (2005) Economics in a full world. Sci Am 293(3):100–107. https://doi.org/10.1038/scientificamerican0905-100

District of North Vancouver (2011) The District of North Vancouver Official Community Plan. Retrieved from https://www.dnv.org/property-and-development/our-official-community-plan-ocp

Dodds F, Donoghue D, Leiva Roesch J (2017) Negotiating the sustainable development goals. Routledge, Abingdon/Oxon/New York

Girardet H (2015) Creating regenerative cities. Routledge, Abingdon/Oxon/New York

Hamstead MP, Quinn MS (2005) Sustainable community development and ecological economics: theoretical convergence and practical implications. Local Environ: Int J Justice Sustain 10(2):141–158. https://doi.org/10.1080/1354983052000330743

[8]Mitacs is a national, not-for-profit organization that builds partnerships to support research and training for industrial and social innovation in Canada. Mitacs Accelerate supports the development of research projects that benefit both graduate students or postdocs and partner organizations. For more information: https://www.mitacs.ca/en/programs/accelerate

Harcourt W (2005) The millennium development goals: a missed opportunity? Development 48(1):1–4. https://doi.org/10.1057/palgrave.development.1100117

Hermans FLP, Haarmann WMF, Dagevos JFLMM (2011) Evaluation of stakeholder participation in monitoring regional sustainable development. Reg Environ Chang 11(4):805–815. https://doi.org/10.1007/s10113-011-0216-y

James P (2015) Urban sustainability in theory and practice | circles of sustainability. Earthscan Publications Ltd, Abingdon/Oxon/New York

Joss S, Cowley R, De Jong M, Müller B, Park BS, Rees WE et al (2015) Tomorrow's City today: prospects for standardising sustainable urban development. University of Westminster, London

Kanie N, Abe N, Iguchi M, Yang J, Kabiri N, Kitamura Y et al (2014) Integration and diffusion in sustainable development goals: learning from the past, looking into the future. Sustainability 6(4):1761–1775. https://doi.org/10.3390/su6041761

Kanuri C, Revi A, Espey J, Kuhle H (2016) Getting started with the SDGs in cities – a guide for stakeholders (UN SDSN). Retrieved from http://unsdsn.org/resources/publications/getting-started-with-the-sdgs-in-cities/

Meadows DH, Meadows DL, Randers J (1992) Beyond the limits: global collapse or a sustainable future. Earthscan Publications Ltd, London

Meerow S, Newell JP, Stults M (2016) Defining urban resilience: a review. Landsc Urban Plan 147:38–49. https://doi.org/10.1016/j.landurbplan.2015.11.011

Mesa N, Edquist M, Espey J (2019) A pathway to sustainable American cities: a guide to implementing the SDGs

Meth P (2013) Millennium development goals and urban informal settlements: unintended consequences. Int Dev Plan Rev 35(1):v–xiii. https://doi.org/10.3828/idpr.2013.1

Moreno Pires S, Magee L, Holden M (2017) Learning from community indicators movements: towards a citizen-powered urban data revolution. Environ Plan C Polit Space 35(7):1304–1323. https://doi.org/10.1177/2399654417691512

Nixon, H. (2016). San José: implementing the UN's sustainable development goals at the local level

Prakash M, Teksoz K, Espey J, Sachs J, Shank M, Schmidt-Traubaub G (2017) Achieving a Sustainable Urban America. Retrieved from http://unsdsn.org/resources/publications/us-cities-sdg-index/

Province of British Columbia (n.d.) Official community plans for local governments. Retrieved August 21, 2019, from https://www2.gov.bc.ca/gov/content/governments/local-governments/planning-land-use/local-government-planning/official-community-plans

Rockström J (2009) A safe operating space for humanity. Nature 461(24 September)

Rockström J, Steffen W, Noone K, Chapin FSI, Nykvist B, de Wit CA et al (2009) Planetary boundaries: exploring the safe operating space for humanity. Ecol Soc 14(2):32. https://doi.org/10.1007/s13398-014-0173-7.2

Roseland M (2000) Sustainable community development: integrating environmental, economic, and social objectives. Prog Plan 54(2):73–132. https://doi.org/10.1016/S0305-9006(00)00003-9
Roseland M (2012) Toward sustainable communities: solutions for citizens and their governments, 4th edn. New Society Publishers, Gabriola Island
Roseland M, Spiliotopoulou M (2016) Converging urban agendas: toward healthy and sustainable communities. Soc Sci 5(3):28. https://doi.org/10.3390/socsci5030028
Roseland M, Spiliotopoulou M (2017) Sustainable community planning and development. In: Abraham MA (ed) Encyclopedia of sustainable technologies. Elsevier, Amsterdam/Oxford/Cambridge, pp 53–61
Roseland M, Spiliotopoulou M (2018) Sustainability in North America: the Canadian experience. In: Brinkmann R, Garren S (eds) The Palgrave handbook of sustainability: case studies and practical solutions. Palgrave Macmillan, Cham, pp 635–652
Ross D (2018) Sustainability planning and assessment: identifying and evaluating community capital in the district of North Vancouver (Unpublished master's thesis/research project). Simon Fraser University, Canada
Ruckstuhl, S., Espey, J., & Rae, L. (2018). The USA sustainable cities initiative: lessons for city-level SDG action
Steffen W, Persson Å, Deutsch L, Zalasiewicz J, Williams M, Richardson K et al (2011) The anthropocene: from global change to planetary stewardship. Ambio 40(7):739–761. https://doi.org/10.1007/s13280-011-0185-x
Steffen W, Richardson K, Rockström J, Cornell SE, Fetzer I, Bennett EM et al (2015) Planetary boundaries: guiding human development on a changing planet. Science 347(6223):1259855. https://doi.org/10.1126/science.1259855
UN DESA (2018) United Nations. World urbanization prospects: the 2018. Revision [Key facts]. Economic and Social Affairs https://doi.org/10.1017/CBO9781107415324.004
United Nations (2015a) Adoption of the Paris Agreement (FCCC/CP/2015/L.9/Rev.1). Paris. Retrieved from http://unfccc.int/resource/docs/2015/cop21/eng/l09r01.pdf
United Nations (2015b) Resolution 70/1. Transforming our world: the 2030 Agenda for Sustainable Development. Retrieved from https://sustainabledevelopment.un.org/post2015/transformingourworld
United Nations (2015c) The millennium development goals report 2015. United Nations. New York. 978-92-1-101320-7
United Nations (2017) New Urban Agenda. Conference on Housing and Sustainable Urban Development (Habitat III) ISBN: 978-92-1-132757-1
Uphoff N (2014) Systems thinking on intensification and sustainability: systems boundaries, processes and dimensions. Curr Opin Environ Sustain 8:89–100. https://doi.org/10.1016/j.cosust.2014.10.010
van Kerkhoff L (2014) Developing integrative research for sustainability science through a complexity principles-based approach. Sustain Sci 9(2):143–155. https://doi.org/10.1007/s11625-013-0203-y
WCED (1987) Report of the world commission on environment and development: our common future (The Brundtland report). https://doi.org/10.1080/07488008808408783
Woodbridge M (2015) From MDGs to SDGs: what are the sustainable development goals? ICLEI – Local Governments for Sustainability
Yin RK (2014) Case study research: design and methods, 5th edn. SAGE, Thousand Oaks

Maria Spiliotopoulou is a Ph.D. candidate at Simon Fraser University's School of Resource and Environmental Management and an instructor in the Sustainable Development Program. Maria's doctoral research aims to advance community sustainability theory by exploring the potential of urban productivity to holistically operationalize sustainable community development. Her goal is to contribute to the global discourse on implementing and assessing local sustainability by proposing new ways to support the achievement of sustainable communities and the UN Sustainable Development Goals. Maria also has an extensive work experience as an environmental consultant in Europe. Her most recent publications, co-authored with Dr. Mark Roseland, include a chapter in the *Palgrave Handbook of Sustainability* (2018) and a chapter in Elsevier's *Encyclopedia of Sustainable Technologies* (2017).

Dr. Mark Roseland is a Director and Professor, School of Community Resources and Development at Arizona State University, and Senior Sustainability Scientist, ASU Global Institute of Sustainability.

Before coming to ASU, Dr. Roseland was a Director of the Centre for Sustainable Development and Professor of Planning in the School of Resource and Environmental Management at Simon Fraser University in Vancouver, Canada. He also worked as Chief City Planner for a municipality in the Metro Vancouver area. Dr. Roseland has been cited by *The Vancouver Sun* as one of Vancouver's top 50 public intellectuals and has received both the SFU Sustainability Network Award for Excellence in Research on Sustainability and the SFU President's Award for Leadership in Sustainability. Dr. Roseland wrote the groundbreaking book *Toward Sustainable Communities: Solutions for Citizens and Their Governments* (New Society Publishers), now in its 4th edition, and founded Pando | Sustainable Communities, a multilingual online network to promote collaboration between researchers and practitioners.

How Local Communities Can Align with United Nations' Sustainable Development Goals: A Santa Cruz County Case Study

6

Susan Brutschy, Jennifer Anderson-Ochoa, James Connery, and Donna R. Maurillo

6.1 Introduction

Now in its 25th year, Santa Cruz County has been collecting data through the Community Assessment Project (CAP) as part of its goal to achieve wellbeing and equity for everyone. Prior to the CAP, social data for the county were collected in a disaggregated way with little "cross-pollination" across sectors and organizations. When the CAP was formed and data began to be collected and organized in a unified manner throughout the county, it presented a significant change to provide a more integrated snapshot of local conditions. More recently, the County has been moving the CAP toward local alignment with the United Nations' 17 Sustainable Development Goals (SDG) established in September 2015.

To help the County achieve wellbeing for all its residents, the United Way of Santa Cruz County (United Way) engaged Applied Survey Research (ASR) as research partner. ASR, a California-based social research organization, has long recognized and shared the story of how communities link their efforts to work collectively toward shared results. ASR has considered the conditions required for communities to work well in a participatory way, to better understand disparities in outcomes, to determine the necessary steps to reduce achievement gaps, and to increase equity in results.

In other words, what outcomes are being realized? What structures and commitments must be in place? What capacity is required for continuous improvement of the ways to achieve desired change? These questions help to position natural integration with the SDGs, which in turn helps communities achieve wellbeing for all.

In this chapter, we will explain the three conditions necessary for applying the international goals of the SDGs locally: commitment to wellbeing for all, population measures of wellbeing, and ability to coordinate action (Fig. 6.1). Then we will provide examples of how Santa Cruz County is aligning those SDGs to its goals by leveraging data from the CAP. Finally, we will show how other communities can do the same.

When considering the three conditions necessary for localizing the SDGs, it is first essential to commit to and work from a position of enabling wellbeing and equity for all. Secondly, communities must link measures of individual and population wellbeing via social connectedness. The third condition captures the ability to align and coordinate action through results-based accountability (RBA) and collective impact models. These approaches yield a great many choices that communities can customize and connect to the

S. Brutschy (✉) · J. Anderson-Ochoa · J. Connery
D. R. Maurillo
Applied Survey Research, Watsonville, CA, USA
e-mail: susan@appliedsurveyresearch.org

D. B. Abraham, S. D. Iyer (eds.), *Promoting the Sustainable Development Goals in North American Cities*, Sustainable Development Goals Series, https://doi.org/10.1007/978-3-030-59173-1_6

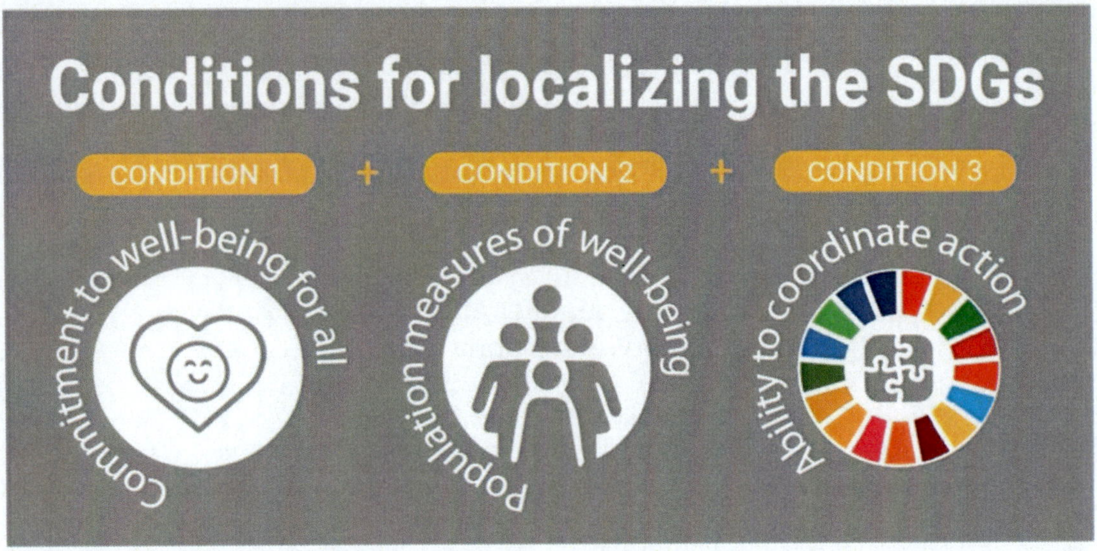

Fig. 6.1 The three conditions necessary for localizing the goals of the SDGs. From Applied Survey Research (2020). Conditions for localizing the SDGs. [Graphic]. (Reprinted with permission)

SDGs in ways that best fit their particular environments.

As these conditions are explored, the Santa Cruz County CAP report and related efforts serve as examples. While the journey to help localize the SDGs will be unique across communities, the conditions considered here offer an orienting framework as a way to achieve universal wellbeing.

6.2 Santa Cruz County Background and the CAP's History

Santa Cruz County is situated alongside the Pacific Ocean on the central coast of California, just south of San Francisco. Of the state's 58 counties, Santa Cruz County ranks in the lower half population-wise, with nearly 276,000 residents (U.S. Census Bureau 2019). Santa Cruz County is composed of four cities—Capitola, Santa Cruz, Scotts Valley, and Watsonville—and its unincorporated areas. While the tech industry of Silicon Valley is just to the north, technology is not a significant part of the local economy. Many local people may be employed in Silicon Valley and may bring home their tech salaries, but the primary drivers of Santa Cruz County's economy include higher education, tourism, and agriculture (Applied Survey Research 2017).

The Community Assessment Project was established in 1994 by leading local entities, most notably the United Way of Santa Cruz County (United Way), with the initial idea coming from Dominican Santa Cruz Hospital. CAP serves the community by guiding plans and processes, examining quality-of-life domains, tracking well-being indicators, and establishing and monitoring community-generated goals. Using these data, Santa Cruz County government, along with other organizations and agencies, seeks to improve the quality of life for all residents. The long-standing success of this community initiative can be attributed to the significant pillars of strong leadership and a commitment to sustainability and innovation.

The CAP showcases how organizations and institutions in Santa Cruz County can work together for the common good of all residents in their respective jurisdictions. It is fundamental to have stable and competent leadership sustaining this effort. Since the CAP's inception, the United Way has been the valuable backbone organization for the project, with ASR serving as its research partner. Other organizations have supported the effort and used this shared dataset, such as local hospitals, law enforcement agencies, educational institutions, and more. Indeed, the CAP steering committee is large, reflecting almost all community initiatives and partnerships.

6.3 Community Conditions Necessary for Applying the UN's SDGs

6.3.1 Condition 1: Commitment to Wellbeing for All

For 25 years, the CAP has orientated its efforts using a five-step community improvement cycle (Fig. 6.2). In so doing, data-informed accountability was embedded in all CAP reports. A commitment to the community to pursue equitable quality of life for all its residents remains prominent. Inclusive engagement is at the forefront of the community improvement cycle, as it is essential for diverse perspectives representing the entire community to be included.

For example, the CAP is governed by a steering committee of more than 30 individuals representing diverse sectors of the county, including nonprofit organizations, government, higher education, and businesses, as well as individual community people. This varied membership holds a collective commitment to equitable wellbeing for all county residents. It is this shared mission that galvanizes their efforts, such as the biennial production of a comprehensive report, including enhancements leveraging past features to support new data, new strategies for engagement of partners, and new ways to share and utilize the findings.

Indeed, CAP's collaborative leadership has helped to position the project at the center of the dedicated work required to establish a countywide Culture of Health. Defined as the circumstances that are cultivated to create and foster equity for all, a Culture of Health seeks to pursue and realize health and wellbeing across various

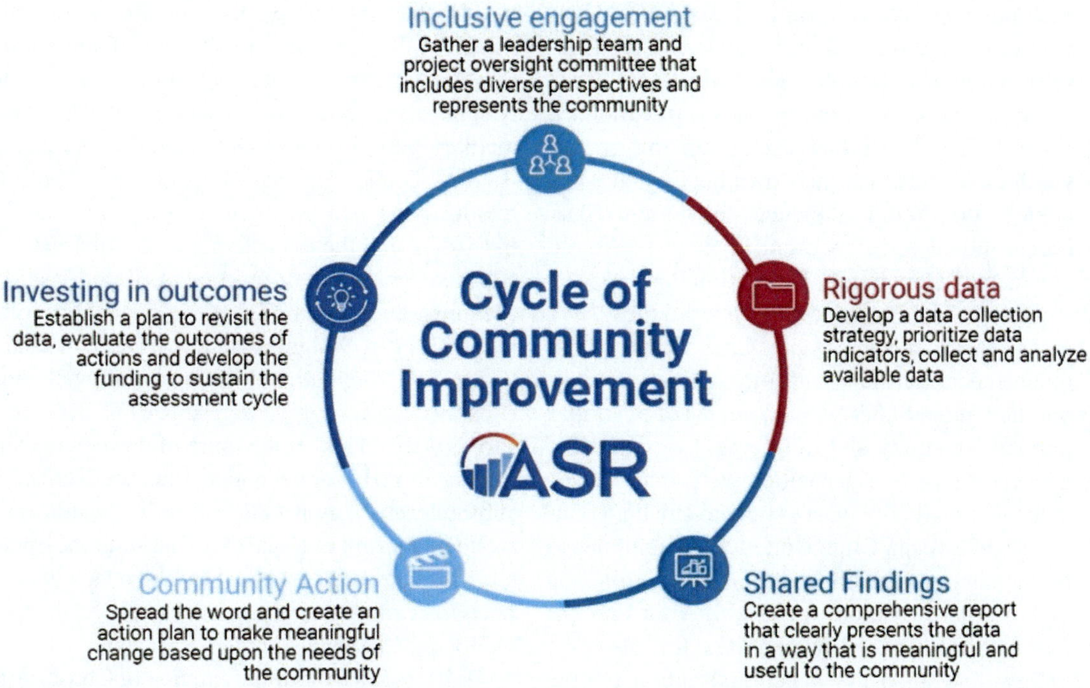

Fig. 6.2 The five-step cycle of community improvement. From Applied Survey Research (2020). Cycle of Community Improvement. [Graphic]. (Reprinted with permission)

social and economic sectors (Evidence for Action n.d.).

While the CAP, under the stewardship of United Way, has received several awards and acknowledgments over its long history, perhaps the most important was inclusion in the 2013 inaugural Culture of Health Award winners. This prize, afforded by the Robert Wood Johnson Foundation (RWJF) and the University of Wisconsin Population Health Institute (UWPHI), recognized Santa Cruz County for centering its efforts in data, ensuring that decision-making was informed, could be measured, and, importantly, could be acted upon (Robert Wood Johnson Foundation 2013). More specifically, acknowledgment was given to the CAP's sustained and continuously improving process to use community assessment data to examine disparities and assets, to ensure community voice is expressed, and to coordinate and motivate initiatives to action through the development of community goals.

In these ways, the CAP put in place a culture of accountability and, notably, supported a nimble system of partnerships that could leverage each other's work. Several partnerships and initiatives were developed in response to CAP findings, including Healthy Kids of Santa Cruz County (health-care program insuring children), *Jóvenes SANOS* (initiative empowering county youth to advocate for their own health and well-being), the Youth Violence Prevention Task Force, and the formerly titled Together for Youth, now named Community Prevention Partners.

Receipt of this recognition elevated the CAP's goals and connected Santa Cruz County with other leaders, partners, and communities throughout the United States who are engaged in similar pursuits for equity and wellbeing. Consequently, new resources and capacities were made available that are usually inaccessible to smaller counties such as Santa Cruz. United Way continues to participate in generating new insights from the growing list of Culture of Health Award recipients from across the United States. It is also providing innovation assistance and action for the CAP and, subsequently, Santa Cruz County to innovate through exploring and adopting new data, indicators, and communication and action strategies (Fig. 6.3). The Culture of Health Alumni network consists of a network of alumni participants from all over the country—a key example of leadership, support, and innovation all working from a shared commitment to wellbeing for everyone.

As the CAP leadership has long maintained the ability to hold itself accountable in striving for a Culture of Health—for wellbeing for all—so do the SDGs hold themselves to a similar standard. This alignment of accountability for intention and action is a foundation for localizing the SDGs, and it sets the stage for adopting the shared measures required to achieve them.

6.3.2 Condition 2: Individual and Population Measures of Well-Being via Social Connectedness

Individual well-being is composed of four domains: physical, mental, emotional, and spiritual. Community well-being, in contrast, is captured in the CAP through five domains: Economic Stability; Education; Social and Community Context; Health and Health Care; and Neighborhood and Built Environment. The intermediary between these two realms of wellbeing is social connectedness, which attests to the role relationships and engagement play both at the individual and the collective levels. The PERMA (Positive emotions, Engagement, Relationships, Meaning, and Accomplishments) model put forth by Dr. Martin Seligman names these five qualities as the foundational elements of wellbeing (Positive Psychology Program 2017). Research supports the idea that the more of these items one has, the more his or her connections are increased, subsequently improving overall health and wellbeing (Kern et al. 2015). That said, the opposite also holds true, as isolation and a lack of connections can negatively impact one's health and wellbeing (Seppala 2014).

In 2016, at a Community Indicator Consortium (CIC) session regarding social connectedness, ASR presented a workshop on how to align these

Fig. 6.3 RWJF Culture of Health Prize Alumni Network, Robert Wood Johnson Foundation (2019). RWJF Culture of Health Prize Winners. [Graphic]. Retrieved from: https://www.countyhealthrankings.org/learn-others/rwjf-culture-health-prize/past-winners

levels of wellbeing and the subsequent implications for the field of wellbeing research and application as a whole (Applied Survey Research 2016). Several takeaways from this session had the most impact, including (1) wellbeing exists in domains beyond those aforementioned; (2) it is important to achieve wellbeing for *all*; and (3) equity has an important role in this conversation.

In effect, this session demonstrated that without these noted components, social connectedness as a bridge between individual and community wellbeing is incomplete and cannot be leveraged or attained. Wellbeing is not a concept that can be measured in a vacuum. It also has a positive impact at individual and community levels by changing the way people work together.

The CAP addresses these concerns and the relationship between individual and collective wellbeing by using common indicators to report population-level data, trend data, and integration of a community survey, which captures the sentiments of a representative sample of the county every two years. CAP secondary data are composed of a public set of population-level quantitative and qualitative measures of wellbeing derived from sources that include the US Census Bureau's American Community Survey. Through the community survey, asset-based measures of subjective and objective aspects of wellbeing are collected and disaggregated, delivering trend data that focuses on both individual and community strengths. Further analysis allows us to explore these data by age, gender, region, housing status, race/ethnicity, and income level. Measures are regularly reviewed to remain comparable with global efforts. This work is supported by associations that Santa Cruz County has cultivated with RWJF and UWPHI's County Health Rankings and its participation in the CIC and the Organisation for Economic Co-operation and Development (OECD), specifically connecting to OECD's work on the Better Life Index. Sustained relationships with these national and international organizations have provided Santa Cruz County the unique opportunity to learn and leverage best practices locally to better explore and understand the alignment between individual and group wellbeing at local, national, and international levels.

As the CAP moves into its 25th year, the natural ways in which the SDGs are present in the report become more formalized, further aligning the goals of both these efforts. Additionally, many of these overlapping measures, as well as those specific to the CAP, offer the necessary information to influence activities intended to strengthen and create connections within the community. This action is at the heart of the final condition and gateway to localizing the SDGs.

6.3.3 Condition 3: Alignment and Ability to Coordinate and Leverage Action

Continuous community improvement necessitates establishing a culture of accountability. This goes beyond intentionality and commitment guided by the tenets of a Culture of Health and implementation of population wellbeing measures informed by community connectedness. This means that resources are applied equitably to population-level community programming efforts in order to coordinate and leverage action.

Early in the CAP process, Santa Cruz County adopted Results-Based Accountability (RBA), developed by Mark Friedman, founder and director of the Fiscal Policy Studies Institute (FPSI) in Santa Fe, New Mexico. This is a way to connect program and population data through a structured way of thinking and acting in order to turn the curve for community change (Friedman 2015). Understanding the work necessary to turn the curve—or to bend data trends toward desired results—is supported through theories of change outlined in community-wide action plans. In fact, the ability to "turn the curve" may be considered one of the most valuable skills to develop if a community is serious about improving outcomes.

A theory of change defines what is required to achieve results, is ideally created with an equity approach, and is predicated upon the notion that working together in specific ways, with certain groups, is essential for community-level change in resource-constrained environments.

Consequently, a theory of change provides coordinated programs and initiatives, a framework by which data can guide the way communities work toward wellbeing for all. This is achieved by manifesting individual and group-level contributions and by monitoring and measuring those contributions transparently. A culture of accountability is created through leveraging RBA alongside a theory of change to solicit responses to the following questions:

1. *How much did we do?* Effectively, this question seeks to understand the number of inputs given toward a particular effort.
2. *How well did we do it?* This question addresses the effort's equity and sustainability.
3. *Is anyone better off?* Ultimately, this question asks whether the desired results were achieved, and if people's lives were improved as a consequence of pursued interventions.

6.4 Santa Cruz County: An Example of Success

For reference, we will explain how Santa Cruz County now aligns its goals with the UN's SDGs. We have included sample results from the 2019 CAP survey.[1] These data will be a valuable resource to guide local governments and other agencies in their quest to improve quality of life for all county residents.

For the 2019 CAP survey, more than 850 county residents age 18 and over were selected by random digit calling. All calls, including those to mobile phones, were done manually to comply with Telephone Consumer Protection Act (TCPA) rules. Selections were adjusted to deliver a fair and equitable representation of local residents. Using Spanish and English, surveyors asked questions in several quality-of-life categories, and the results were tabulated and compared with the data from previous years. This helped to provide a "progress map" that revealed trends in various classifications, such as employment sta-

[1]The complete report is available at www.appliedsurvey-research.org

tus, food security, access to health care, educational attainment, and other relevant categories. As with past CAP reports, positive trends will be further supported, while negative or neutral trends will be examined for opportunities to improve outcomes.

6.5 Specific SDGs Directly Connected to CAP Social Determinants of Health (SDOH) Components

The primary purpose of the United Nations' SDGs is to balance the three dimensions of sustainable development—economic growth, environmental sustainability, and social inclusion. The CAP report has recently incorporated these SDGs while continuing to address the SDOH inequities and resource disparities that were uncovered in Santa Cruz County during previous surveys. As such, the CAP data are now organized in relationship to the five key areas of SDOH: Economic Stability; Education; Social and Community Context; Health and Health Care; and Neighborhood and Built Environment.

Three SDGs—No Poverty; Zero Hunger; and Decent Work and Economic Growth—were aligned to the Economic Stability SDOH and its accompanying goals:

- Goal 1: By 2020, reduce the winter unemployment rate by one-half percent, creating 725 new winter jobs in Santa Cruz County.
- Goal 2: By 2020, increase the housing stock by 1000 units in Santa Cruz County.

ECONOMIC STABILITY

NO POVERTY
End poverty in all its forms everywhere. Economic growth must be inclusive to provide sustainable jobs and promote equality.

ZERO HUNGER
End hunger, achieve food security and improved nutrition, and promote sustainable agriculture.

DECENT WORK AND ECONOMIC GROWTH
Promote sustained, inclusive and sustainable economic growth, full and productive employment, and decent work for all.

Agriculture is one of the two largest employment sectors in Santa Cruz County, and the county has a high concentration of jobs in this sector when compared with the rest of California.[2]

Tourism also ranks as one of the top employers and revenue-producing industries in Santa Cruz County, breaking the $1 billion mark in travel-related spending for the first time in 2017 and for a second time in 2018.[3] Visitor dollars help to support Santa Cruz County by providing business and tax revenues, which contribute to local employment; open space, beaches, and parks; and locally owned small businesses.[4]

The 2019 CAP report shows that the gap has continued to narrow between the available workforce and total employment (141,700 workforce vs. 97,600 jobs in 2010; 144,900 workforce vs. 113,800 jobs in 2018). Tourism jobs have also increased (7580 in 2010 vs. 11,403 in 2018). Therefore, the County is on a positive track and should continue with current activities.

[2]Santa Cruz County. (2014). *Economic Vitality Study*.

[3]Dean Runyan Associates, Inc. (2019). *California Travel Impacts 2010-2018p*.

[4]Visit Santa Cruz County (VSCC). *Tourism Facts*. Accessed on July 26, 2016, from http://www.santacruzca.org/partners/tourism-facts.php

The County learned through the 2019 CAP report that residents are not feeling better off financially now than they were a year ago—40.5% feel better off, which is a decline of 1.6% over the past 12 years. However, between 2009 and 2017, respondents reporting annual family income of at least $75,000 increased by 27%.

According to the California Housing Partnership (CHP), Santa Cruz County must create 11,873 more affordable rental homes, equivalent to 27% of the existing rental stock, to meet affordable housing needs.[5]

EDUCATION

QUALITY EDUCATION

Ensure inclusive and equitable quality education, and promote lifelong learning opportunities for all. This is the foundation for improving people's lives and creating sustainable development.

The CAP report revealed that median home prices have increased significantly since 2012—from $426,000 to $743,000. This means that only 17.3% of homes are affordable to median-income families, versus 53.8% in 2010. (The US median home price in 2019 was $260,000, with 61.4% affordable for median-income families.)

Therefore, the County has learned that, while incomes and employment opportunities may be increasing, the ability to afford housing in the local market has declined. Perhaps this is why fewer people believe they are better off financially than they were a year ago even though they are earning more.

One SDG—Quality Education—was aligned to the Education SDOH and its accompanying goals:

- Goal 1: By 2020, all students will be fully connected and engaged with their school community, and they will see their school as a welcoming, essential, and safe place.
- Goal 2: By 2020, all students will have broader access to courses and enrichment activities, including visual and performing arts, career technical education, and digital technology.
- Goal 3: By 2020, all students will be provided sufficient behavior, health, and counseling services to succeed in their chosen educational and career pathways.

High quality, developmentally appropriate early childhood education (ECE) produces positive effects on children's cognitive and social development.[6] Moreover, studies of the costs and long-term benefits of these ECE programs have consistently found substantial savings derived over decades, such as reduced need for remedial and special education, reduced incarceration rates, and lower rates of teen pregnancy. Analyses of the costs and benefits of ECE show a 13% per year return on investment. Additionally, research has found that high quality and reliable child care increases employee productivity and improves the bottom line for business.[7,8]

[5]California Housing Partnership. (2018 September). *Santa Cruz County's Housing Emergency and Proposed Solutions.*

[6]NAEYC. A Call for Excellence in Early Childhood Education. Accessed on October 2, 2019, from https://www.naeyc.org/policy/excellence

[7]University of California, Berkeley, Center for Labor Research and Education. *Economic Impacts of Early Care and Education in California.*

[8]Heckman, James, The Life-cycle Benefits of an Influential early Childhood Program. Accessed on October 17, 2017, from https://heckmanequation.org/the-heckman-equation/

However, according to the 2019 CAP report, 41.1% of parents say that finding affordable child care is a "somewhat serious" or "very serious" problem. Here is another area where relevant partners are collaborating to solve these issues.

Studies in the United States and internationally show that the more developmental assets young people acquire, the better their chances of succeeding in school and becoming happy, healthy, and contributing members of their communities.[9] Importantly, they are less likely to engage in high-risk behaviors.[10]

The 2019 CAP report revealed that 50% of seventh-grade students say they have someone at school who encourages them and expects the best. By eleventh grade, that number drops to 41%. This is an opportunity for education-related initiatives to support students, teachers, and staff developing stronger relationships.

America's schools must meet the educational needs of an increasingly diverse student population.[11] English Learners (ELs) are the fastest-growing student population within the United States, and a large achievement gap exists between ELs and their non-EL classmates.[12]

In Santa Cruz County, the number of ELs has dropped from 28.9% in 2010 to 25.8% in 2018. During the same time, special education enrollment has increased from 11.3% to 13.2%. All of these figures are higher than the average for the state of California.

High school graduation rates have remained at about 81.7% between 2011 and 2018, but they did reach a high of 87.4% in 2014. In all but one year, the rate has exceeded that of California.

Without a college degree, children born in the bottom income quintile have a 45% chance of remaining there as adults. With a degree, they have less than a 20% chance of staying in the bottom quintile and a roughly equal chance of ending up in any of the higher income quintiles.[13]

The CAP survey reveals that enrollment in Cabrillo College (the local community college) dropped from 13,825 to 11,648 between 2012 and 2018. However, enrollment at the University of California, Santa Cruz, increased from 17,404 to 19,700.

Graduation rates are at around 90% or more for most schools, although in reality, all students should be achieving a high school diploma or its equivalent. They also should be moving on to college, vocational schools, or career training so they can have healthy and successful lives as adults.

As one example, Pajaro Valley Unified School District in Santa Cruz County has been addressing the literacy challenge head-on. It instituted *Paso a Paso* (Step by Step), which offers a variety of reading and alphabet assignments for children from toddlerhood up to third grade. Parents can download the assignments, given in English and Spanish, and work with their children on fun activities such as tracing letters, singing The Alphabet Song, reading stories, or engaging in related endeavors. At the completion of each set of tasks, the child receives a certificate.

One SDG—Good Health and Well-Being—was aligned to the Health and Health Care SDOH and its accompanying goals:

- Goal 1: By 2020, all Santa Cruz County residents will have a regular source of primary care and integrated behavioral health services with a focus on:
  - Decreasing disparities
  - Decreasing reliance on Emergency Rooms as a regular source of health care
  - Increasing access to mental health and substance use disorder treatment
- Goal 2: By 2020, obesity in Santa Cruz County will be reduced by 10%.

[9]Search Institute. Accessed on August 2, 2016, from http://www.search-institute.org/research/developmental-assets

[10]Search Institute. Accessed on August 2, 2016, from http://www.search-institute.org/research/developmental-assets

[11]U.S. Department of Education. 2010. *Diverse Learners.*

[12]U.S. Department of Education. 2010. *Diverse Learners.*

[13]U.S. Department of the Treasury with the U.S. Department of Education. 2012. *The Economics of Higher Education.*

HEALTH AND HEALTH CARE

GOOD HEALTH AND WELL-BEING

Ensure healthy lives and promote well-being for all ages, which is essential for sustainable development.

Individuals without a dependable source of health care have more difficulties obtaining needed care, receive fewer preventive health services, are more likely to wait until their conditions worsen before seeking treatment, and are more likely to require hospitalization compared with those who have a dependable source of health care.[14,15] Children's access to primary health care is especially important to monitor healthy growth and development[16] and to prevent everyday illnesses from progressing into more serious problems. Children with a usual source of care are more likely to utilize preventive services and to have better health outcomes and fewer disparities overall.[17]

CAP survey respondents in 2019 (88%) reported at similar percentages to previous years that in the past 12 months they were able to receive the health care that they needed, although Whites were more likely to receive it than Latinos. About 60% received care in a regular doctor's office, while about 25% received it at a community clinic or hospital. The remainder had no reliable source.

A lack of health insurance coverage is a barrier to accessing health services. Families and individuals without health insurance coverage often have unmet health needs, receive fewer preventive services, suffer delays in receiving appropriate care, and experience more hospitalizations.[18]

Children who have health insurance learn better in school, miss fewer days of school, are more likely to have a regular source of primary care, and are less likely to be hospitalized for conditions that could have been treated by a primary care physician.[19]

Pregnancy can provide an opportunity to identify existing health risks in women and to prevent future health problems for women and their children. The risk of pregnancy-related complications and maternal and infant mortality can be reduced by increasing access to quality care.[20] Regular prenatal care reduces the risk of adverse birth outcomes, including preterm and low birth weight babies.[21]

In Santa Cruz County, about 87% of pregnant women receive prenatal care during the first tri-

[14] U.S. Department of Health and Human Services, Agency for Healthcare Research and Quality. (2011). National Healthcare Disparities and Quality Report. Washington, DC.

[15] Billings, J., Bidman, A.B., Grumbach, K., et al. (1995). Preventable hospitalizations and access to health care. Journal of American Medical Association, 274(4): 305–311.

[16] ChildStats.gov. *America's Children: Key National Indicators of Well-Being*. Usual Source of Health Care. 2015. http://www.childstats.gov/americaschildren/care2.asp. Accessed October 21, 2015.

[17] U.S. Department of Health and Human Services, Health Resources and Services Administration, Maternal and Child Health Bureau. Child Health USA 2014. Rockville, Maryland: U.S. Department of Health and Human Services, 2014.

[18] U.S. Department of Health and Human Services. (2011). Healthy People 2020 objectives. Retrieved from http://healthypeople.gov/2020/topicsobjectives2020/overview.aspx?topicid=1

[19] Bernstein J, Chollet D, Peterson S. *How Does Insurance Coverage Improve Health Outcomes?*. ISSUE BRIEF 2010. Available at: http://www.mathematica-mpr.com/~/media/publications/pdfs/health/reformhealthcare_ib1.pdf. Accessed October 21, 2015.

[20] U.S. Department of Health and Human Services, Healthy People 2020. Healthy People 2020. Accessed on August 9, 2016, from https://www.healthypeople.gov/2020/topics-objectives/topic/maternal-infant-and-child-health

[21] What is PedNSS/PNSS? *Centers for Disease Control and Prevention* 2011. Available at: http://www.cdc.gov/pedness/what_is/pnss_health_indicators.htm. Accessed October 2015.

mester. That's an increase of four percentage points from 2012. Teen mothers are less likely to receive this care than mothers in older age groups.

Teen parents and their children are often at greater risk for experiencing negative short- and long-term consequences in the areas of health, school, and economic success, as compared with parents who wait to have children.[22] Research from the National Campaign to Prevent Teen and Unplanned Pregnancy links teen pregnancy to preterm births, low birth weight, and a host of social issues, including poverty, responsible fatherhood, and overall wellbeing.[23]

There is good news from the 2019 CAP report. Girls from 15–17 years are less likely to become pregnant now than in 2012 (12.4% vs. 4.2%). The same is true for girls 18–19 years (26.6% vs. 10.6%).

When asked about their mental health, the percentage of CAP overall survey respondents who had felt so sad or hopeless almost every day for two weeks or more in a row that they stopped doing some usual activities increased from 6.0% in 2017 to 8.0% in 2019. In 2019, Latino survey respondents were slightly more likely than White respondents to feel this way—8.0% and 4.2%, respectively. Adolescent hospitalizations for emotional issues also increased. For every 1000 children from 5–14 years old, 2.2 were hospitalized in 2016, along with 9.4 of every 1000 aged 15–19 years.

Healthful diets and healthy body weights reduce the risk of chronic diseases and promote optimum health.[24] Efforts to change diet and weight should address individual behaviors, as well as the policies and environments that support these behaviors in settings such as schools, worksites, health care organizations, and communities. For example, having healthful food available and affordable in food retail and food service settings allows people to make more healthful food choices.[25]

To address some of these issues, the United Way established *Jóvenes SANOS*, a youth advocacy and leadership program seeking to empower, educate, and raise awareness about childhood obesity within the community. The purpose is to shift the local culture by encouraging businesses to offer healthful food choices and encouraging people to increase their physical activity.

For example, *Jóvenes SANOS* has prompted corner markets to offer more fresh, non-processed foods; it encouraged restaurants to include more healthful choices on their menus; and it requested that the local Metro service include at least 50% healthful items in its vending machines.

Survey respondents overall have consistently described their general health as Excellent, Very Good, or Good (83.7% in 2019) at percentages exceeding the Healthy People 2020 target of 79.8%. In 2019, Latino survey respondents (80.6%) were significantly less likely than White survey respondents (85.7%) to describe their health as Very Good or Excellent. Over the past 10 years, there has been a steady increase in the number of overall survey respondents who were obese and overweight based on BMI calculation (61.7% in 2019). Latino survey respondents (76.3%) were more likely than White survey respondents (53.7%) to be obese or overweight in 2019. Clearly, the organizations across the county must educate the population and motivate them to reduce their weight and follow better health practices.

On a positive note, children in Santa Cruz County are much less likely than their statewide counterparts to drink one or more sugar-

[22] Kaye K, Stewart Ng A. TEEN CHILDBEARING, EDUCATION, AND ECONOMIC WELLBEING. *Why It Matters: Teen Childbearing, Education, and Economic Wellbeing* 2012. Available at: https://thenationalcampaign.org/sites/default/files/resource-primary-download/childbearing-education-economicwellbeing.pdf. Accessed October 2015.

[23] National Campaign to Prevent Teen and Unplanned Pregnancy. *Why it matters: Teen pregnancy*. 2012. Accessed November 2015 from: http://www.thenationalcampaign.org/why-it-matters/wim_teens.aspx

[24] U.S. Department of Health and Human Services, Healthy People 2020. Healthy People 2020. Accessed December 2015 from https://www.healthypeople.gov/2020/topics-objectives/topic/nutrition-and-weight-status

[25] Centers for Disease Control and Prevention. Overweight & Obesity. *Healthy Food Environments*. Accessed December 2015 from http://www.cdc.gov/obesity/strategies/healthy-food-env.html

sweetened beverages per day. In 2011, the rate was 30% in Santa Cruz County versus 41% in California. By 2015, those numbers were 17% and 43%, respectively.

According to the 2019 CAP report, binge drinking has increased in Santa Cruz County, moving from a low of 12.2% in 2009 to a high of 21.2% in 2019. Likewise, adolescents who ever have had an alcoholic drink have moved from a low of 18.6% in 2014 to a level of 54.3% in 2019. However, that has decreased from a high of 65.1% in 2015.

Two SDGs—Reduced Inequalities and Sustainable Cities and Communities—were aligned to the Social and Community Context SDOH and its accompanying goals:

- Goal 1: By 2020, more Santa Cruz County residents will build meaningful social bridges across differences in age, race, ethnicity, class, and culture.
- Goal 2: By 2020, schools and communities will be safe, supportive, and engaging places for children, youth, and families.
- Goal 3: By 2020, more Santa Cruz County residents will feel empowered to experience and pursue long-term quality of life.

SOCIAL AND COMMUNITY CONTEXT

REDUCED INEQUALITIES

Reduce inequality within and among countries. Policies must be universal in principle while meeting the needs of disadvantaged and marginalized populations.

SUSTAINABLE CITIES AND COMMUNITIES

Make cities and human settlements inclusive, safe, resilient and sustainable. Communities must provide opportunities for all, with access to basic services, energy, housing, transportation, and more.

Although health is one of the important domains of overall quality of life, there are other domains as well, including jobs, housing, schools, the neighborhood, aspects of culture, values, and spirituality.[26] Focusing on quality of life as an outcome can bridge boundaries between disciplines and between social, mental, and medical services.

Over one-third (35.5%) of overall survey respondents believe that the cost of living/housing in Santa Cruz County takes away from their quality of life, followed by homelessness (22.3%). Since 2013, cost of living/housing has risen from the fifth-highest concern (13.0%) among survey respondents to the first (35.5%), a 173% increase. The life satisfaction ladder captures a snapshot of wellbeing by asking survey respondents to rank where they land on a scale of 0–10, with 10 being their best possible life and 0 their worst possible life. One-half (50.8%) of survey respondents rated themselves highly on this ladder (rungs 8–10), 41.6% rated themselves a 5–7, and 7.6% responded 0–4.

Homelessness is usually the result of the cumulative impact of several factors, rather than a single cause. The scarcity of affordable housing in the United States, particularly in more urban areas where homelessness is more prevalent, is a major structural barrier to acquiring or maintaining housing.[27]

[26] Centers for Disease Control and Prevention. Health-Related Quality of Life. Accessed September 2016 from http://www.cdc.gov/hrqol/concept.htm

[27] National Alliance to End Homelessness. Snapshot of Homelessness. Accessed September 2016 from http://www.endhomelessness.org/pages/snapshot_of_homelessness

Nationally, 552,830 people were homeless on a given night in the United States in January 2018.[28] Of that number, 33% were people in families, and 67% were individuals.[29] The State of California had the highest rate of persons experiencing homelessness in unsheltered situations, with a rate of nearly 70% of the total homeless population.[30] Locally, Santa Cruz County has one of the largest concentrations of people experiencing homelessness in unsheltered situations (78%).[31]

That said, the number of homeless people in Santa Cruz County has fallen from a high of 3,789 in 2007 to 2,167 in 2019. However, homelessness is difficult to eradicate because many live under the radar, some prefer living this way, others have complex situations that may include physical and mental health issues, and other factors.

Solving these issues requires creative approaches. So, Santa Cruz County has instituted the Youth Homeless Demonstration Project with a grant from the federal government. This program targets the estimated 600 unaccompanied homeless youth, offering a drop-in center, host homes, rapid rehousing, a program for disabled youth or those with complex problems, and other programs—including those intended to prevent youth from becoming homeless.

Civic engagement refers to individual and collective actions designed to identify and address issues of public concern.[32] Studies show that volunteers become emotionally connected to the communities they serve, and they sustain community involvement after volunteering.[33]

Overall survey respondents reported in 2019 that they had participated at lower levels in various civic engagement activities in the last 12 months than during the previous 2017 survey year, including voting, attending public meetings, and communication with a local politician. How do we re-engage those people? The positive result would be a more equitable connection with their local neighborhoods and communities.

Four SDGs—Peace, Justice, and Strong Institutions; Climate Action; Life Below the Water; and Life on Land—were aligned to the Neighborhood and Built Environment SDOH and its accompanying goals:

Public Safety

- Goal 1: By 2020, the juvenile crime rate will be reduced by 10% through the use of culturally responsive evidence-based strategies that promote positive interaction and reduce conflict with public safety officials.
- Goal 2: By 2020, there will be a 20% reduction in youth reporting gang involvement, resulting in a 10% reduction of gang-related criminal activity.
- Goal 3: By 2020, there will be a 10% decrease in arrests or citations of individuals with chronic SUD/COD through the increase of on-demand treatment for adults with such disorders.
- Goal 4: By 2020, the violent crime rate of 18- to 25-year-olds will be reduced by 10% through the use of targeted gang involvement intervention strategies, including restorative

[28] The U.S. Department of Housing and Urban Development. (2019). *The 2018 Annual Homeless Assessment Report to Congress.*

[29] The U.S. Department of Housing and Urban Development. (2019). *The 2018 Annual Homeless Assessment Report to Congress.*

[30] The U.S. Department of Housing and Urban Development. (2019). *The 2018 Annual Homeless Assessment Report to Congress.*

[31] The U.S. Department of Housing and Urban Development. (2019). *The 2018 Annual Homeless Assessment Report to Congress.*

[32] American Psychological Association. Civic Engagement. Accessed September 2016 from http://www.apa.org/education/undergrad/civic-engagement.aspx

[33] Gergen, Christopher. (2012, April 17). The Benefits of Civic Engagement for Tomorrow's Leaders. Accessed September 2016 from https://www.whitehouse.gov/blog/2012/04/17/benefits-civic-engagement-tomorrows-leaders

practices, street outreach, and alternatives to adult gang involvement.

Natural Environment

- Goal 1: By 2020, residential per capita water use will be sustained at or under 2013 baseline levels through 2020.
- Goal 2: By 2020, 5% of homes in Santa Cruz County will have a solar electric or hot water system.
- Goal 3: By 2020, stewardship actions for our waters will be increased by 10%.
- Goal 4: By 2020, 50 miles of urban bike and multi-use trails will be constructed within Santa Cruz County to decrease traffic, increase active transportation, and connect urban areas to open spaces.

NEIGHBORHOOD AND BUILT ENVIRONMENT

PEACE, JUSTICE & STRONG INSTITUTIONS
Promote peaceful and inclusive societies, provide access to justice for all, and build effective, accountable institutions at all levels.

CLIMATE ACTION
Take urgent action to combat climate change and its negative impacts because this is a global challenge that affects everyone.

LIFE BELOW THE WATER
Conserve and sustainably use the oceans, seas, and marine resources because careful management of this essential global resource is key to a sustainable future.

LIFE ON LAND
Sustainably manage forests, combat desertification, halt and reverse land degradation, and halt biodiversity loss.

Unsafe neighborhoods are associated with high rates of infant mortality and low birth weight, juvenile delinquency, high school dropout, child abuse and neglect, and poor motor and social development among preschool children.[34] Conversely, children who live in highly supportive neighborhoods have positive outcomes, such as stronger connections with family, peers, and community, and greater participation in out-of-school programs, volunteering, and religious services.[35]

More than one-half (58.2%) of survey respondents said they felt Very Safe in their neighborhoods, while 29.8% of survey respondents were Very Concerned about crime in Santa Cruz County. Renter survey respondents were significantly more likely than homeowner survey respondents to answer Very Concerned or Somewhat Concerned about violent crime and gangs in their neighborhoods in 2019. But 86.3% of respondents said that law enforcement personnel were Somewhat Trustworthy or Very Trustworthy.

Gangs operate in cities of all sizes throughout California and are responsible for much of the crime in the state.[36] Research suggests that a comprehensive approach to gangs involving pre-

[34]Child Trends Data Bank. Neighborhood Safety. Accessed September 2016 from http://www.childtrends.org/?indicators=neighborhood-safety

[35]Child Trends Data Bank. Neighborhood Safety. Accessed September 2016 from http://www.childtrends.org/?indicators=neighborhood-safety

[36]California Department of Justice, Division of Law Enforcement, Bureau of Investigation and Intelligence. (2010). Organized Crime in California. Retrieved from: http://oag.ca.gov/sites/all/files/agweb/pdfs/publications/org_crime2010.pdf

vention, intervention, and suppression efforts works better than suppression efforts alone.[37]

Youth who have been involved with the juvenile justice system are at increased risk of substance abuse, injury, and negative educational impacts. Many factors have been noted as contributing to crime among youth, including poverty, exposure to violence, maltreatment, substance abuse, and mental illness.[38] Youth who have spent time in detention are more likely to engage in criminal behavior as adults and experience increased rates of attempted suicide and other mental health disorders.[39]

To address these particular issues, local groups established the Youth Violence Prevention Task Force as a way to build better relationships and improve trust among young people, adults, and law enforcement. Recently, several facilitated dialogs were held in each law enforcement jurisdiction so youth, adults, and police could engage in honest discourse about their challenges, opportunities, local experiences, and other issues that may prevent collaboration on community safety. As expected, some discussions were difficult, but they were successful in helping each group better understand the others.

The 2019 CAP report shows that the crime rate in Santa Cruz County has decreased from 36.9 per 1000 residents in 2012 to 31.2 per 1000 residents in 2018. Although this is encouraging, the rate is still slightly higher than California as a whole (28.5 per 1000 residents). However, the homicide rate for the County has dropped from 4.5 per 100,000 people in 2012 to 2.3 per 100,000 people in 2017. This is lower than the California rate of 5.2 per 100,000.

Regarding the Natural Environment, the CAP reports that water use per person has increased between 2015 and 2018. In addition, county residents have increased the amount of trash per person between 2011 and 2017. California likewise has increased in the same time period. It's apparent that residents need more education about conserving resources, especially in a state that has a reputation for environmentalism.

During the years between 2012 and 2018, county-wide transit ridership decreased from 5,465,542 annual riders to 5,048,512. Meanwhile, 69% of workers in 2018 drove alone to work, a dip of 1.1% from 2012. Again, this is an area for further investigation regarding the factors leading to these changes and what can be done to transform transportation habits into those that are more environmentally responsible.

Poor air quality is harmful to people and can cause a variety of environmental problems, such as effects on wildlife, ozone depletion, and global climate change.[40] Some groups of people are especially sensitive to poor air quality, including those with asthma, heart disease, and COPD (long-term lung disease).[41] Outdoor air quality has improved since the 1990s, but many air quality problems persist. Ground-level ozone, the main part of smog, and particle pollution are two of the many threats to air quality and public health in the United States.[42]

6.6 A Detailed Example: Improving the Lives of Children and Youth

Grown out of a response to CAP findings, the County's Youth Violence Prevention Task Force (YVPTF) was formed in late 2012 to assess and

[37] Howell, J. C. (2007). Menacing or mimicking? Realities of youth gangs. Juvenile and Family Court Journal, 58(2), 39–50. Retrieved from: http://www.nationalgangcenter.gov/Content/Documents/Menacing-or-Mimicking.pdf

[38] Juvenile Arrests Summary – Kidsdata.org. *Kidsdata.org*. Available at: http://www.kidsdata.org/topic/30/juvenile-arrests/summary#jump=why-important. Accessed November 23, 2015.

[39] Prevent Juvenile Delinquency. *Policy for Results* 2015. Available at: http://www.policyforresults.org/youth/prevent-juvenile-delinquency. Accessed November 23, 2015.

[40] Commonwealth of Massachusetts, Executive Office of Energy and Environmental Affairs, Department of Environmental Protection. Accessed September 2016 from http://www.mass.gov/eea/docs/dep/air/aq/health-and-env-effects-air-pollutions.pdf

[41] Centers for Disease Control and Prevention. Air Quality. http://www.cdc.gov/air/

[42] Centers for Disease Control and Prevention. Air Quality. http://www.cdc.gov/air/

address the needs and assets of Santa Cruz County relative to youth violence (United Way of Santa Cruz County n.d.). The YVPTF utilized CAP data and leveraged the RBA format to develop its strategic plan, which included evidence-based strategies to help turn the curve on youth violence in the county. The strategic plan continues to ensure that both organizational partners and community members have a shared understanding of their roles in supporting the community's youth, both now and in the future. The long-term commitment of the YVPTF's members is also emphasized. Prior to establishing the YVPTF, Santa Cruz County had first gained notoriety for its use of RBA in successfully turning the curve on youth substance abuse. The initiative, Together for Youth (now Community Prevention Partners), implemented strategies that influenced data trends, bending them toward desired results (Friedman 2015).

Undeniably, change is ever a constant, and the CAP's efforts to innovate have hardly stood still as the years have advanced. In fall 2018, with continued leadership from United Way, the CAP unveiled a new product, the Children and Youth Well-being Spotlight (CYWB). This document has focused on the most salient indicators impacting the wellbeing of the community's young people. Remaining mindful of the importance of social connectedness, these measures are organized under four headings: economic wellbeing, education, health, and family and community. While the CAP took steps in past reporting years to more closely position its efforts beyond the community level, such as alignment with the Social Determinants of Health, in similar form, the CYWB made specific connections to the aspirations of the SDGs with its release.

Explicit integration of the SDGs into this and future efforts illustrate how Santa Cruz County continues to iteratively improve itself and its partnerships. Moreover, disparities were called out in the CYWB data. This shined a light on existing inequities within the county, with the intended goals of inciting interest, spurring engagement, and inspiring action. By focusing on those children and youth who were not at grade level or who did not earn their high school diplomas, the report emphasized the necessity of creating goals that would change trajectories for more vulnerable populations.

The life course of the CAP project has supported all the conditions necessary to localize the SDGs, providing solid framework for the CYWB to build upon. Indeed, this framework allowed the leadership of the CYWB to quickly mobilize under shared purpose, developing both an ambitious report and a dynamic, web-based tool to help improve the lives of all young people in the county (Fig. 6.4).

Now Santa Cruz County is moving ahead with the next phase—a natural integration with the United Nations' SDGs, which help to achieve wellbeing for all people. Adopted by Member States of the United Nations in 2015, the 17 Sustainable Development Goals set forth an ambitious agenda outlining the path to wellbeing for all by 2030, importantly recognizing the interconnectedness of strategies necessary for accomplishing these goals (United Nations n.d.).

By connecting the work of the CYWB to the SDGs, the ways in which real and tangible results are realized locally can be connected back to global efforts of the UN's SDGs. Normalizing the SDGs through named inclusion in CAP products encourages familiarity with the goals themselves as well as with the greater motivation for their initial establishment—to build a better world for everyone by 2030.

6.7 Conclusion

This article demonstrates how the United Nations' Sustainable Development Goals can be leveraged to help communities around the world set workable and equitable goals to help improve quality of life for their residents. Santa Cruz County has been progressing for 25 years, thanks to its Community Assessment Project, which provides a clear overview of community quality of life, the successes in making positive changes, and the direction to improve less-than-ideal conditions. By incorporating the UN's SDGs into its existing Social Determinants of Health, the

Fig. 6.4 The 2018 Children and Youth Well-being Spotlight was promoted through report and online formats. From Applied Survey Research (2018). 2018 Children and Youth Well-being Spotlight. [Graphic]. (Reprinted with permission)

County has now become integrated with a global movement toward health justice for everyone.

6.8 Generalizations

In general, it is most beneficial for communities to localize the Sustainable Development Goals by:

1. *Connecting to larger initiatives that focus on effective action, such as Social Determinants of Health (SDOH) and Organisation for Economic Co-operation and Development (OECD)*—When communities associate their goals with those of regional and global initiatives, it allows them to become part of a momentous collaboration targeting common issues that affect people everywhere. Each community interconnects to form a network stretching across the globe, solving humanity's problems through cooperative involvement.
2. *Approaching the entire project from a paradigm of wellbeing, equity, and health justice*—The primary goal is for communities to develop conditions that support a positive quality of life for all residents. Nearly all else is secondary.
3. *Building on successful strategies*—When a community reflects on its successes, it is already moving in a positive direction. What strategies have been effective in the past to improve community health and wellbeing? What conditions helped to make those strategies successful? What lessons were learned from any missteps? How can those lessons be applied here? These kinds of questions help a community move quickly in a positive direction with minimal blunders.
4. *Using a community assessment based on resident data, combined with objective and/or secondary data*—The most reliable and beneficial data will come from the people themselves—their experiences, their sense of what is important for them, their perceptions of where essential processes have failed or succeeded, and so on. However, although it is valuable, individual perception is not always reality. Therefore, objective data also must be included to help determine whether personal perceptions reflect reality for the greater population.
5. *Maintaining a sense of urgency*—To delay is to lose valuable time that could be used for improving community wellbeing. Therefore, plan to act immediately or at least within a reasonable timeframe. It may be necessary to cull the activity list if it becomes a source of delay. Achieving a few goals in the near term is preferable to delaying action because of an inability to achieve a lengthy list of goals.

6.9 Recommendations

Communities wishing to leverage the UN's SDGs to help improve quality of life for their residents and for the global population are advised to follow a particular course of action. These recommendations may be followed "as is," or they may be adapted to specific needs within the community.

The first step is to establish the conditions necessary for community involvement. These conditions are detailed earlier in this paper: commitment to well-being for all; population measures of well-being; and ability to coordinate action. Details may be adapted to fit the needs of specific communities, but the process should remain essentially intact.

The next step is to align your work within the framework of the SDGs by selecting those that are most relevant to your situations or that are most manageable and attainable. For example, large municipalities may have more resources to implement a more complete list, while smaller towns may be limited by lesser funding and fewer committee participants. The goal is to develop a plan that fits a community's particular needs and resources. Highly visible and decisive results from these activities will help the community expand into other SDGs. Conversely, taking on too many goals at the onset will stall progress and discourage further participation. Not all SDGs need be implemented at once.

In Santa Cruz County, the United Way took on a leadership role for developing the CAP, with participation from many other local agencies and organizations—hospitals, educational institutions, social service agencies, private individuals, nonprofit organizations, and other entities that saw mutual benefit from this type of cooperation. As a group, they helped determine the types of data they would require to achieve the Social Determinants of Health and, later, the SDGs. Participating organizations pooled their resources, according to their abilities, to help fund the CAP. Because the project budget has been responsibly managed, the CAP has remained sustainable throughout the years, becoming the longest-running project of its type.

Next, conducting statistically valid, professionally administered surveys will provide the necessary data for a snapshot of current quality-of-life conditions and—with each successive survey—a comparison with past conditions. These surveys should generate data that can be applied to a community's chosen SDGs, enabling a clear illustration of current status and a point from which to formulate improvement goals.

Note that it is important to engage professionals in creating the surveys to ensure that they are worded in a neutral fashion, so the questions do not compel the respondent to reply in a particular way. Professional survey organizations also will (or should) be involved with an institutional review board to ensure that respondents can feel secure in the knowledge that honest responses will not put them in jeopardy, legally or otherwise. It's recommended that professionals also conduct the surveys for the most valid results. However, if funds are limited, volunteers may perform this task—but only if they are *properly trained and supervised* in the fine points of survey taking.

Resulting data should be summed in a public report that is also available to the news media. This provides valuable information not only to citizens but also to agencies and organizations that are in a position to improve any negative outcomes and to further support positive outcomes.

And finally, communities should track progress toward meeting their chosen SDG goals. Again, those goals will be different for every community, as evidenced by the way the City of Santa Cruz has instituted its own Health in All Policies, which addresses issues specific to the local community.

This means it will be necessary to follow up with additional surveys at regular intervals. This may be annually or biennially, based on local resources. In any event, they should not be scheduled at lengthy intervals because new data are necessary to track progress and allow processes to be adjusted, if necessary.

By following the United Nations' SDGs, more communities, more nations, and ultimately the entire globe can enjoy true equity and health justice, along with the benefits of economic stability, quality education, improved health and health care, more livable communities, and a cleaner and more sustainable environment.

References

Applied Survey Research (2016, September 27) Community connectedness and wellbeing at the local, regional, and (inter)national levels (PowerPoint Presentation)

Applied Survey Research (2017, November) Santa Cruz County Community Assessment Project, Year 23. Retrieved from: https://www.appliedsurveyresearch.org/scccap

Evidence for Action (n.d.) What is a culture of health? Retrieved from Evidence for Action: https://www.evidenceforaction.org/what-culture-health

Friedman M (2015) Trying hard is not good enough: how to produce measurable improvements for customers and communities. PARSE Publishing. Retrieved from http://resultsaccountability.com/about/what-is-results-based-accountability/

Kern ML, Waters LE, Adler A, White MA (2015) A multidimensional approach to measuring well-being in students: Application of the PERMA framework. J Posit Psychol, 10(3):262–271. https://doi.org/10.1080/17439760.2014.936962

Positive Psychology Program (2017, February 24) The PERMA model: your scientific theory of happiness. Retrieved from: https://positivepsychologyprogram.com/perma-model/

Robert Wood Johnson Foundation (2013, February 13) Santa Cruz County, CA: 2013 RWJF Culture of Health Prize. Retrieved from Robert Wood Johnson Foundation: https://www.rwjf.org/en/library/articles-and-news/2013/02/santa-cruz-county%2D%2Dcalifornia%2D%2D2013-roadmaps-to-health-prize.html

Seppala E (2014, May 8) Connectedness & Health: the science of social connection. Retrieved from Stanford Medicine: The Center for Compassion and Altruism Research and Education: http://ccare.stanford.edu/uncategorized/connectedness-health-the-science-of-social-connection-infographic/

U.S. Census Bureau (2019) American community survey: community facts – Santa Cruz County. Retrieved from: https://factfinder.census.gov/faces/nav/jsf/pages/community_facts.xhtml

United Nations (n.d.) Sustainable Development Goals. Retrieved from United Nations: https://sustainabledevelopment.un.org/?menu=1300

United Way of Santa Cruz County (n.d.) Retrieved from Santa Cruz County's Youth Violence Prevention Task Force: http://www.sccyvpt.org/

Susan Brutschy is president of Applied Survey Research (ASR), which she co-founded in 1981 with Dane Archer, PhD, sociology professor at the University of California, Santa Cruz (UCSC). Since then, she has led hundreds of evaluations, assessments, and strategic planning processes to uncover and leverage data about community-wide quality of life in the United States and internationally. ASR has earned awards from the Brookings Institution and the Association for Applied and Clinical Sociology. Ms. Brutschy graduated from UCSC with highest honors in sociology. She also received awards for her research in designing and implementing quantitative assessments of community opinions for UCSC's Social Research Unit.

Jennifer Anderson-Ochoa is project manager at Applied Survey Research (ASR) and has worked on assessments in community health, adult and youth homelessness, and child health. Her recent projects include the Dignity Health Dominican Hospital Community Health Needs Assessment, the San Luis Obispo Community Assessment Project, and the Stanislaus Youth and Family Well-Being Spotlight. Previously, she directed two family resource centers in Santa Cruz County. Ms. Anderson-Ochoa worked for 23 years in HIV/AIDS prevention in the United States and in sub-Saharan Africa. She received her Masters' Degree in international administration, with a concentration in cross-cultural communication and training, from the School for International Training in Brattleboro, VT. She received her BA in sociology from George Washington University.

James Connery is project manager at Applied Survey Research (ASR). Since 2010, he has contributed to Community Assessment Projects throughout California and Alaska by collecting data and collaborating with members of communities to develop data that otherwise would not be available. In addition, he has worked on needs and assets reports of children ages birth through five years in nine regions of Arizona. He has also contributed to homeless censuses and surveys for several counties. James is a dedicated San Francisco Bay Area sports fan and loves to see how statistics are increasingly used as an integral part of sports performance evaluation. Mr. Connery graduated from the University of Arizona with a Bachelor of Arts in history. He earned his MBA from California State University, Monterey Bay.

Donna R. Maurillo is communications director for Applied Survey Research. She has a lifelong career in public relations and corporate messaging, primarily for health care, technology, public policy, and higher education. Most recently, she consulted to Cisco Systems, and she was director of Technology Transfer for the Mineta Transportation Institute, a research organization affiliated with San Jose State University (SJSU). She has managed projects with state and federal agencies, the Chinese and South Korean governments, and Cabinet-level offices. She earned her BA in liberal arts with honors from the University of California, Santa Cruz, and her MSc in transportation management from SJSU.

7 Top-Down and Bottom-Up Approaches to the SDG Monitoring Challenge

Jessica Espey

7.1 Introduction

The Sustainable Development Goals (SDGs), universally adopted by the world's governments in 2015, aim to set a framework for action on economic development, social inclusion, and environmental sustainability. In the USA, the engagement of local government leaders in SDG implementation and associated monitoring is crucial as 85% of the domestic population lives in cities and their surrounding metropolitan areas.[1] These cities are centers of economic enterprise and innovation. In 2017 the ten largest metropolitan areas generated $6.8 trillion in economic value, surpassing the output of the sum of 37 US states.[2] But they are also responsible for much of the country's waste and environmental destruction, including more than 80% of the country's CO_2 emissions.[3] It is cities in the USA that will make or break sustainable development for the country.

In support of city-level action on the SDGs, the UN Sustainable Development Solutions Network (SDSN) has been working with US cities since late 2014 exploring ways of localizing and implementing the global goals. Central to all of these discussions has been data, data which can tell city representatives their starting point, can support them to set realistic benchmarks between now and the 2030 deadline, and can help track their progress. Indeed, data has been such a foundational aspect of all of the local SDG implementation conversations SDSN has had across US cities (including in Atlanta, Baltimore, Boston, Los Angeles, Minneapolis, Milwaukee, New York, Orlando, San Jose, and South Bend) that in 2016 SDSN's urban and data programs launched a crossover initiative called the Local Data Action project which aims to create a library of case studies and technical knowledge documenting how to engage with and monitor the SDGs at city and regional levels. This has been a useful technical exercise which has showcased different methods and approaches for integrating

[1]SDSN's calculations. See Espey, Jessica, Dahmm, Hayden, and Laurie Manderino (2018) Leaving No US Cities Behind: The U.S. Cities SDG Index, Issue 2018, New York: UNSDSN. Available at: http://unsdsn.org/wp-content/uploads/2018/06/US-Cities-Index-Report.pdf

[2]USCM (2018) US Metro Economies: Economic Growth and Full Employment, Annual GMP Report, Prepared for The United States Conference of Mayors and The Council on Metro Economies and the New American City by HIS Markit. Available at: http://www.usmayors.org/wp-content/uploads/2018/06/Metro-Economies-GMP-June-2018.pdf

[3]Jones, C., and D. Kammen, (2014) Spatial Distribution of U.S. Household Carbon Footprints Reveals Suburbanization Undermines Greenhouse Gas Benefits of Urban Population Density, Environmental Science & Technology, Vol 48, Issue 2, Pages 895–902.

J. Espey (✉)
Sustainable Development Solutions Network, New York, NY, USA
e-mail: jessica.espey@unsdsn.org

D. B. Abraham, S. D. Iyer (eds.), *Promoting the Sustainable Development Goals in North American Cities*, Sustainable Development Goals Series, https://doi.org/10.1007/978-3-030-59173-1_7

the SDGs in local planning and monitoring local progress; however local processes and data sources vary considerably, and so the local data generated is not directly comparable.

To support a more active comparison of progress across the USA and to encourage more cities to take up the SDGs, SDSN also pioneered a US Cities SDG Index.[4] The index, first launched in 2017, repeated in 2018 and 2019, aims to provide a set of comparable metrics across the 100 largest metropolitan areas in the USA, which show overarching progress on the Sustainable Development Goals. The SDG Index enables us to see which US cities and regions are faring well or performing badly on specific goals. The Index consists of 44 indicators spanning 15 of the 17 SDGs. Goal 14 on Life Below Water and Goal 17 on Partnership for the Goals are excluded since they do not apply to many US cities and/or data are insufficient.

Both of these approaches to local SDG monitoring have benefits and limitations. A national index is advantageous in that it enables active comparison, can help generate support for the goals, can show areas that are underserved, and can help direct federal political attention and investment. Conversely a local, bottom-up approach to monitoring enables cities to utilize existing data resources and to map the alignment of their current policies and planning to the goals; it can also foster community engagement and buy in. This chapter critically evaluates the benefits and limitations of both approaches and makes recommendations on how cities should approach the challenge of local monitoring of the SDGs. It draws heavily upon learning from the SDSN's USA Sustainable Cities Initiative, the preparation of the SDSN's US Cities Index 2017 and 2018, and SDSN TReNDS' Local Data Action project.[5]

7.2 A National Index for US Cities

7.2.1 Methods

The US Cities SDG Indices (2017, 2018, 2019), prepared by the SDSN, provide a portrait of sustainable development at the local level for the 100 most populous metropolitan areas in the USA. In all studies, the metropolitan statistical area (MSA) is used as the geographic unit instead of the nuclear city, because more comparable data are available at this level. Additionally, many of the SDG challenges translate most naturally onto the interconnected metropolitan region rather than individual jurisdictions within the MSAs. Nonetheless, the term "city" is used interchangeably with MSA to make the report more accessible and policy intuitive. Taken together the 100 MSAs within the index are home to 66% of the US population. The SDG Index enables us to see how US individual cities are performing on specific indicators for each goal. The 2018 index, which is a composite index, was calculated using 44 indicators spanning 15 of the 17 SDGs.

For each goal in the US Cities SDG Index, indicators that evaluate aspects of sustainable development have been identified, for which data are readily available and are consistently collected across the country. These indicators map closely to the set of global SDG indicators proposed by the UN's Inter-Agency and Expert Group on SDG Indicators but were selected primarily based on their relevance to the US context and their availability.

Although the index was calculated for 3 different years (2017, 2018, 2019), the results are not directly comparable. Improvements were made between versions to strengthen the meth-

[4]Espey, Jessica, Dahmm, Hayden, and Laurie Manderino (2018) Leaving No US Cities Behind: The U.S. Cities SDG Index, Issue 2018, New York: UNSDSN. Available at: http://unsdsn.org/wp-content/uploads/2018/06/US-Cities-Index-Report.pdf and Prakash, Mihir, Teksoz, Katerina, Espey, Jessica, Sachs, Jeffrey, Shank, Michael and Guido Schmidt-Traub (2017) Achieving a Sustainable Urban America, The U.S. Cities SDG Index 2017, New York: UNSDSN. Available at: http://unsdsn.org/wp-content/uploads/2017/08/US-Cities-SDG-Index-2017.pdf

[5]For more information, visit: http://unsdsn.org/what-we-do/solution-initiatives/usa-sustainable-cities-initiative-usa-sci/ and https://www.sdsntrends.org/local-data-action

odology, as well as to add some new indicators such as the food insecurity rate, infant birth weight, the percentage of 3–4 year-olds enrolled in school, and the percentage of businesses owned by women—all of which are crucial measures for understanding equality of opportunity across the USA. Overall, annual calculation of indices found that all US cities featured in this Index perform poorly on one or more goals and 60% of the cities studied are less than half way there, highlighting widespread sustainable development challenges such as environmental degradation, access to vital infrastructure, and social disparities, among others.

The reports are intended to serve as a tool for US cities to track their progress over time, relative to an international standard of sustainable development. It is also hoped that the indices will enable cities to identify peers struggling with similar challenges and help facilitate a national dialogue on how to accelerate progress.

7.2.2 Impact and Lessons

Since the publication of the US Cities SDG Index, there are some discernible lessons, relating to the utility of the index for trends analysis and political engagement, as well as some limitations, resulting from data availability and the utility of the data collected for local application.

7.3 Identifying Aggregate Trends Across the Country

In the 2017 edition of the US Cities SDG Index, a few striking results such as higher incidences of child *poverty* and acute *racial inequalities* across cities were highlighted. In the 2018 edition, these social inequalities were examined in more detail to better understand not only in which cities and MSAs the poorest and most marginalized live but also how social inequalities might be deepening deprivations within cities.

7.3.1 The First Major Finding: Poverty

Through disaggregation and correlation analysis, results show child poverty rates in nearly all MSAs are larger than poverty rates for the rest of the population (except Provo-Orem, UT). Nine MSAs have child poverty rates that are more than 50% greater than the overall poverty rate in the given MSA. Cape Coral-Fort Myers, Florida (FL), has a child poverty rate 67% greater than that of the overall poverty rate in that MSA. McAllen-Edinburg-Mission, Texas (TX), has the highest child poverty rate in the country at 44.7%, compared to a whole of population poverty rate of 32.8% for the same region. Early poverty is associated with negative outcomes later in life. The analysis performed corroborates this, as child poverty has been found to be correlated with youth being out of education or employment across the 100 MSAs in the sample (see Fig. 7.1).

7.3.2 The Second Major Finding: Racial Inequalities

US cities experience deep racial disparities. In 57 MSAs in the 2018 index sample, the poverty rate among non-whites is at least twice that of whites, and in 6 MSAs, it is over 3 times the rate for whites (see Fig. 7.2). Similarly, non-white unemployment rates are at least 50% greater than that for whites in 73 MSAs and are twice as large as that for whites in 34 MSAs (Espey et al. 2018).

These findings serve to demonstrate the multidimensional nature of poverty in US cities and how different forms of inequality and deprivation can keep people trapped in cycles of poverty and poor health. The analysis found correlations between high poverty levels among non-whites, infant mortality, and other acute health concerns such as food insecurity, obesity, and deaths from heart attack, reinforcing other literature and panel studies that suggest poverty can affect the entire life cycle.

The compilation of national indices on the SDGs, drawing upon federally comparable indicators, has proven exceptionally helpful for

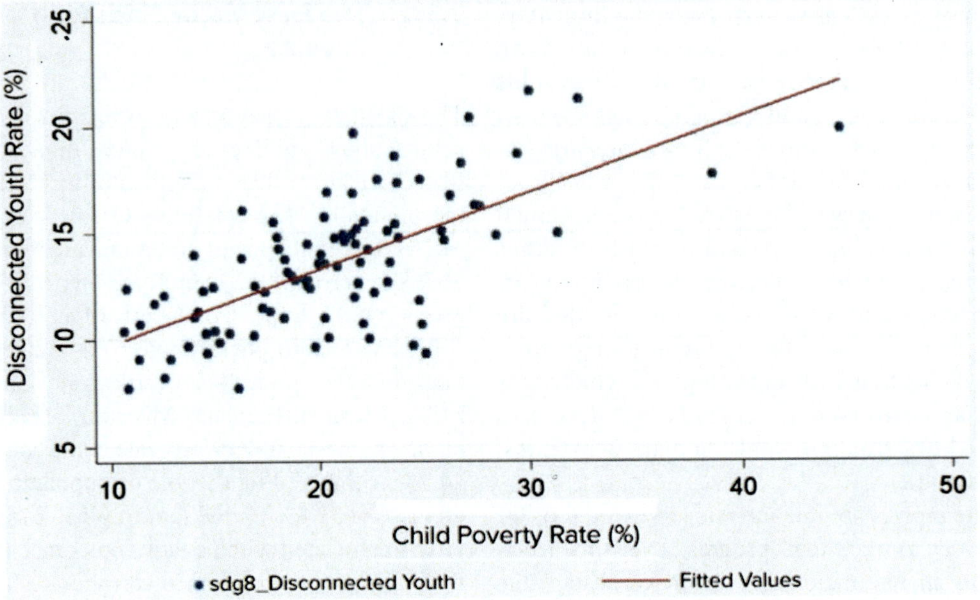

Fig. 7.1 Correlation of child poverty and disconnected youth in US cities. (Source: Espey et al. 2018, p. 28)

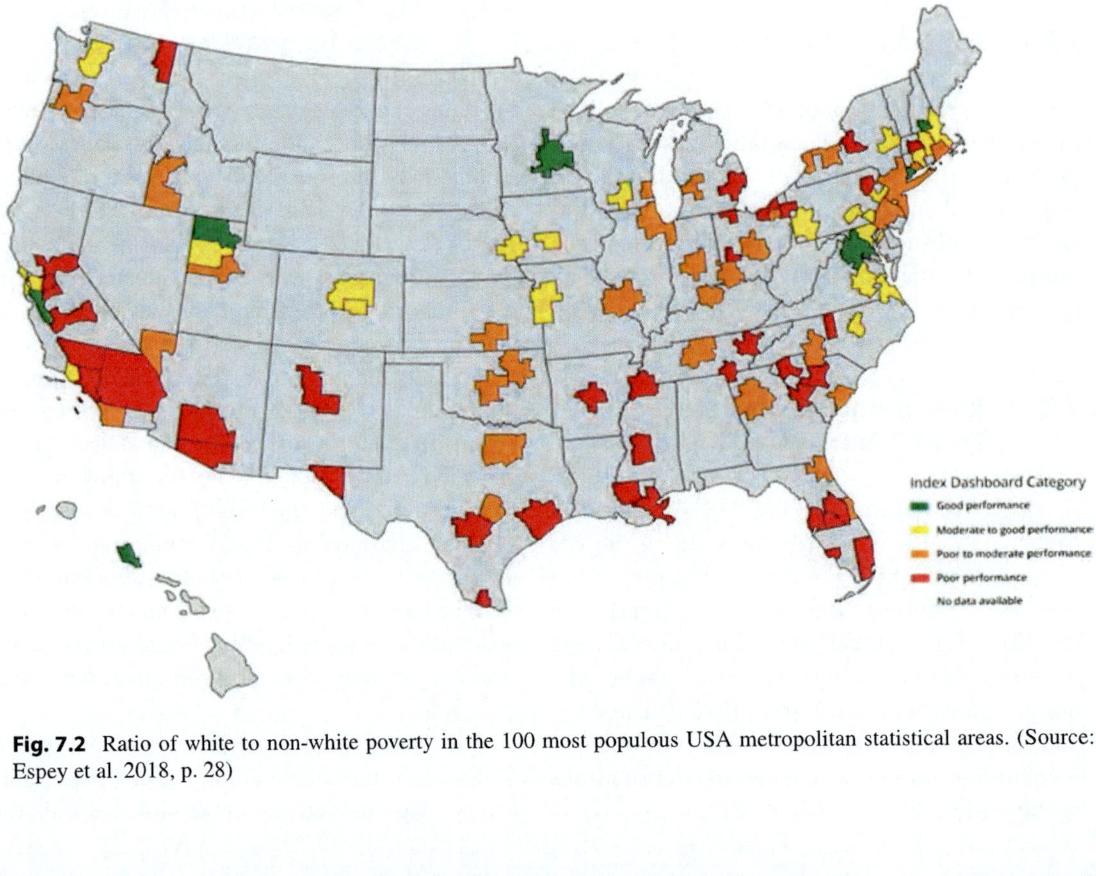

Fig. 7.2 Ratio of white to non-white poverty in the 100 most populous USA metropolitan statistical areas. (Source: Espey et al. 2018, p. 28)

understanding common challenges and trends across US cities, around which mayors and local government leaders can mobilize and share lessons. Trend analysis such as this also hints at policy interventions, such as targeted policies early in the life cycle, which can help disrupt inherited disadvantage and the intergenerational transfer of poverty.

However, the utility of this kind of analysis for policy development is limited by the fact that only federally comparable data could be used, which is often geographically limited, meaning it cannot be disaggregated within cities, by blocks, or specific grid streets. This kind of disaggregation is often essential for local governments looking to provide cost-effective targeted interventions to particular communities.

7.4 Political Engagement

The US Cities SDG Index reports are intended to be a technical resource but also an advocacy tool. Even in the past year, the index report has helped to foster interest in the SDGs among mayors and other local government leaders on the relevance and utility of the SDG framework, for example, through discussion sessions at meetings of the US Conference of Mayors and by encouraging shared learning among well-performing and struggling cities.[6] San Jose and Los Angeles, for example, who both perform well on the 2017 and 2018 index reports have used their rankings to produce articles and host local seminars on the relevance of the SDGs with the active engagement of their city leadership, thereby cementing their political commitment to the SDG agenda.[7] The reports themselves have also garnered considerable media interest from outlets such as USA News, Bloomberg, Vice, Fast Company, the Seattle Times, and the Boston Globe.[8] This has helped to spur interest from city officials in areas not currently mainstreaming or adopting the SDG framework and thereby kick-starting conversations on SDG implementation, for example, in Boston, MA; Cambridge, MA; Orlando, FL; and Atlanta, GA.

Given that the indices derive a large number of the measures from federal data sources, the national-level cities index has encouraged positive exchange with the US federal government regarding open data and governance. Focus is on how to integrate these metrics into the federal government dashboard (https://sdg.data.gov) and use the index to help track the status of sustainable development across its cities and MSAs.

While this kind of media attention and federal-level engagement has helped to focus attention on the SDGs and has piqued the interest of new localities, it has not encouraged a substantive discussion about how we approach local monitoring or policy implementation. Instead, most cities SDSN has engaged with have referred to the index reports as a tool to kick-start a conversation on shared challenges and collaborative working across cities.

7.5 Data Availability

A major constraint in preparing the 2017, 2018 and 2019 indexes was data availability. Many crucial sustainable development indicators had to be excluded as the data were either unavailable, were not standardized across cities, or had limited coverage. It was therefore necessary to use data at the level of the broader metropolitan statistical area, which opened up data sources like

[6] http://unsdsn.org/news/2018/01/29/us-mayors-inspiring-local-change-through-global-sdgs/ [Last accessed 27.10.2019] http://unsdsn.org/news/2019/03/01/sdsn-publishes-new-guide-for-us-cities/ [Last accessed 27.10.2019] https://www.sdgcompacts.org/news/2018/1/25/us-conference-of-mayors-sdgs [Last accessed 27.10.2019].

[7] https://www.fastcompany.com/40451569/how-u-s-cities-stack-up-on-the-sustainable-development-goals [Last accessed 27.10.2019] https://grist.org/article/which-american-cities-are-the-most-sustainable/ [Last accessed 27.10.2019]. https://www.lamayor.org/mayor-garcetti-announces-partnership-occidental-college-advance-sustainable-development-goals [Last accessed 27.10.2019].

[8] http://unsdsn.org/news/2018/06/26/media-wrap-up-of-the-2018-u-s-cities-sdg-index-report/ [Last accessed 27.10.2019].

the Census and its associated American Community Survey.

Positively, MSAs provide a more holistic picture of local sustainable development as they typically represent a large central city and adjacent areas of regional influence. This provides a larger representation of an urban settlement. However, even at the level of the MSA, data availability was limited. For example, to provide an indicator for *Goal 7: Ensure access to affordable, reliable, sustainable and modern energy for all*, state-level data had to be drawn upon and values assigned to the MSA. Therefore, the data lacks a certain geographic precision, as data on the source of energy at the city level was unavailable.

Other indicators such as "number of homes with rooftop solar panels" or "local investments in renewable energy" were explored, but no consistent or standard metric was available. Similarly, "carbon emissions per capita" is the only indicator under *Goal 13: Take urgent action to combat climate change and its impacts*. This variable comes from a nongovernmental source—Berkeley University's Cool Climate Institute.[9] Indicators that measured urban disaster risk management and resiliency planning were pursued, but no standard measures across enough MSAs were available.

Perhaps most shockingly, a number of fundamental life-saving indicators were also not available or were severely limited, such as maternal mortality data, including teenage births, mental health, and drug usage. There are a number of reasons for this, including a lack of open data policies, as well as the absence of federal standards for the compilation of many nationwide health indicators. As such, many of the health indicators collected sub-nationally are done so using locally defined methodologies which are, in some instances, irreconcilable:

> As of 2018, only seven of the 100 MSAs had accessible data on maternal mortality, while more than half of the teenage birth data provided at the MSA level have poor population coverage. A similar problem was experienced with regards to infant mortality data; although 97 MSAs have some available data on infant mortality, only 61 MSAs have data covering 75% or more of the population.[10]

The limited availability of open, comparable data across MSAs, and cities, poses a severe challenge for the federal government if it is to judiciously allocate resources and direct policies to support the SDGs and for citizens to monitor change and hold their leaders to account. Furthermore, it compromises regional approaches, across cities and states, on shared challenges such as waste management, population migration, drug policing, and other issues which are central to all US cities and regions.

7.6 Ground-Up SDG Localization

7.6.1 Methods

In September 2015, the Sustainable Development Solutions Network (SDSN) partnered with leading academic institutions through the USA Sustainable Cities Initiative (USA-SCI) to pilot processes for long-term strategies on the Sustainable Development Goals (SDGs) in three US cities: New York, San José, and Baltimore. The foundation of the pilot cities' SDG strategy process was "start with where we are" and as such to look at existing city plans and programs, as well as data to see how the city was fairing on the 17 SDG goals and associated 149 targets. Across all three cities, residents and officials agreed that the SDG indicators and data provided a common language for strategy building, helping to structure coherent discussions about coordinated city initiatives in order to meet the goals by 2030 and beyond. The first activity undertaken in all three cities was therefore to map existing

[9]For more information, visit their website. Available at: http://coolclimate.berkeley.edu/index [Last accessed March 2019].

[10]Espey, Jessica, Dahmm, Hayden, and Laurie Manderino (2018) Leaving No US Cities Behind: The U.S. Cities SDG Index, Issue 2018, New York: UNSDSN. Available at: http://unsdsn.org/wp-content/uploads/2018/06/US-Cities-Index-Report.pdf (p. 33).

data sources to the SDG indicators (for which an SDG Mapping Worksheet was developed[11]). Thereafter city policies and plans of relevance would be mapped to the targets.

Given the centrality placed upon data in these initial city strategy discussions, in 2016 SDSN went a step further, launching a Local Data Action (LDA) project exclusively focused on different approaches to local SDG monitoring. The Local Data Action project, a joint endeavor by SDSN Cities and SDSN TReNDS, did not present a particular model but instead aimed to create a library of case studies and technical knowledge documenting how global cities and localities were currently engaging with and monitoring the SDGs. Knowledge was curated locally, in consultation with city staff, technical partners, and other stakeholders. As of 2019, SDSN has worked with nine partners representing cities, regions, and networks of cities from around the world.[12] The group explored themes related to (1) indicator localization (how to tailor the global indicators to the subnational context and identify additional local indicators to promote SDG action and achievement); (2) data platforms (identifying data dashboard models to provide easy-to-use granular data on SDG dimensions); (3) the use of third-party data (filling sub-national data system gaps with third-party data, such as citizen-generated or telecommunications data); and (4) national to local data integration (specifically focusing on methods for aligning and integrating national and subnational SDG reporting systems).[13] The case studies were structured around five key questions: (1) What are the most pressing SDG-related problems? (2) What was the identified data solution? (3) What was the building process, including the various steps involved? (4) What are the strengths and weaknesses of the approach, as discussed with local stakeholders? (5) How replicable is the model in other contexts?

7.6.2 Impact and Lessons

As a result of these programs, SDSN has documented emerging practices on local SDG monitoring in four US cities; Baltimore, San Jose, New York, and Los Angeles.

Four years since SDSN's localization studies began, we have identified a number of insights from across cities and regions working to monitor and achieve the SDGs.

7.7 Local SDG Monitoring Efforts Gain Most Traction When Aligned with Existing City Planning and Measurement Frameworks

Nearly all of the cities and regions studied noted that it was hard to pique local political interest unless the SDG framework was presented as an additive framework that would support and improve upon existing plans and commitments, including the priorities articulated by the mayor or local government leader in their election manifestos. As such, local partners within academic institutions and city stakeholders recommended mapping SDG indicators onto existing local strategies and monitoring tools, for example, in San Jose, local academic partners undertaking SDG mapping looked at how the SDGs might support the city's new 15-year Climate Smart San Jose sustainability plan. In part because of this work, the city placed a strong emphasis on people and quality of life, as well as environmental outcomes, in the final strategy launched in 2017.[14] This approach was mirrored on the other side of the world, in Patiala, India, where the SDG strategy was developed around the stated priorities of

[11]Available at: http://unsdsn.org/what-we-do/solution-initiatives/usa-sustainable-cities-initiative-usa-sci/

[12]Aruba, Belo Horizonte in Brazil, a network of municipalities in Colombia, Patiala in India, LA in the USA, Bristol in the UK (all 2018), Baltimore in USA (2017), California Bay Area in the USA (2017), and a network of municipalities in Brazil (2017). Visit: https://www.sdsntrends.org/local-data-action

[13]For more information, visit: https://www.sdsntrends.org/local-data-action

[14]For more on the Climate Smart San Jose plan, visit http://www.sanjoseca.gov/climatesmartsanjose

the city's leadership, which aligned with SDGs relating to health (3), water and sanitation (6), industry and infrastructure (9), sustainable cities (11), climate change (13), and good governance (16). This simple connect-the-dots approach was found to reduce any skepticism and improve buy-in from local officials.

In the case of LA, where the mayor has played a leadership role in promoting the SDGs, the local research team developed a list of proposed local SDG indicators that aligned with LA's Sustainable City pLAn. The team aimed to propose a set of targets and associated indicators that would enable a more coordinated government effort to achieve the SDGs.

The advantage of aligning SDG planning and monitoring exercises with existing local strategies and plans is that it can encourage better cross-governmental coordination and minimize excessive, cumbersome local monitoring. But unless a clear commitment is also made to progressively meet the other "missing" goal areas, it risks being a "pick-and-choose" approach which may jeopardize the integrated and indivisible nature of the SDGs.

7.8 The SDGs and the Official Indicators Provide a Common Language, to Encourage Coordination and Where Possible Active Comparison

Across the cities studied, local stakeholders praised the utility of the SDG indicator framework (the set of 240 indicators recommended by Inter-Agency and Expert Group on Sustainable Development Goal Indicators – IAEG-SDG[15]) for providing a common language—one which enabled diverse city stakeholders to talk about their objectives in reference to specific metrics and outcomes. Using a data-informed approach also helped ensure that conversations about priorities and targets (including target thresholds) were evidence-based and locally relevant.

All of the cities studied drew upon the set of IAEG-SDG indicators to some degree in their initial discussions, either just for inspiration when designing their own locally relevant indicators or as the basis for their monitoring framework. On a few dimensions, such as CO_2 emissions and urban sprawl, the discussions have consequently spawned cross-city discussions on methodological alignment (e.g., in the sidelines of the recent Winter US Conference of Mayors meeting), which may eventually enable active cross-city comparison. However, it was pointed out by a number of cities, such as LA, USA, and Bristol, UK, that global indicators are not always directly relevant and appropriate for a city context and the city's jurisdiction may limit its ability to affect achievement. For example, target levels (e.g., national versus sub-national), geographic context (e.g., coastal versus landlocked), and various data constraints have implications for how cities utilize the official indicators and structure city-level SDG monitoring. Therefore, it is imperative that local stakeholders critically analyze the relevance of the local IAEG-SDG indicators and then work together to craft a functional set of local indicators that can better support local policies and planning.

7.9 The Necessity to Develop Additional Local SDG Measures

In all of the cities studied, stakeholders felt the need to tailor the IAEG-SDG indicators or significantly add to them to better reflect local priorities and conditions. For example, in Baltimore, stakeholder discussions on appropriate measurement indicators for SDG 1 turned to the topic of causes of poverty in the city. Stakeholders concluded that "liquid asset poverty," households with at least 3 months of accessible cash, is a strong indication that a household may not be resilient to shocks such as layoffs in an economic recession, illness of the household breadwinner, or property damage in an environmental disaster.

[15] https://unstats.un.org/sdgs/iaeg-sdgs/ See also the e-handbook on SDG indicators, available at https://unstats.un.org/wiki/display/SDGeHandbook/Home

Consequently, the group determined that a measure of liquid asset poverty should be included as an indicator to track the city's progress in achieving SDG 1, over a more basic measure of income poverty. There are clear benefits to tailoring the IAEG indicators to make them more locally relevant and useful for policy purposes; however, the wide variation in SDG indicators being used across cities, and very different approaches being utilized to identify new indicators or proxies for SDG outcomes in different cities, presents challenges when seeking to review local, regional, or aggregate national progress and/or to use local data to complement national SDG monitoring efforts.

7.10 Local SDG Monitoring Is Encouraging the Use of New Data Sources and Is Supporting the Push for Open Data

Across all of the cities studied, acute data gaps were identified where both federal and local data were missing for key SDG dimensions, for example, timely, disaggregated measures of maternal mortality. City stakeholders expressed their eagerness to fill these gaps as soon as possible, rather than waiting for the production of additional official statistics from the National Statistical Office or local government. City stakeholders would utilize new methodologies and external partnerships, even though these processes are costly and time-consuming. In San Jose, for example, Stanford University provided a prototype dashboard to the city, of the kinds of block-level analysis that could be done using third-party data on vehicle miles traveled and workers commuting time. Specifically, they used LEHD Origin-Destination Employment Statistics (LODES) and the Google Maps Directions API to estimate the vehicle miles traveled (VMT) by both residents and workers who drive alone to and from individual block groups.[16] Discussions on new data sources and methods were active in all of the cities studied, though few had actively started curating third-party data over time, suggesting the challenges of finalizing methods and brokering fair, secure, and sustainable agreements with third-party data providers.[17]

In all of the cities studied, local stakeholders were eager to set up SDG data monitoring mechanisms and platforms which would enable easy tracking of SDG progress. Academic partners in San José and Baltimore are currently in the process of researching and setting up SDG data systems for the cities, which are open source, align with existing datasets, and provide user-friendly visualization tools for policy-makers and public citizens. Additionally, in New York, groups like Measure of America are looking at how to expand their current open-source city dashboards to also map the city's OneNYC and SDG indicators. All three cities are considering methods for integrating these with the US's national reporting platform for the SDGs. In every case, city stakeholders expressed the necessity to make the data dashboards open source and readily accessible for government policy-makers and planners, as well as local residents so they might use the dashboard to track progress within their communities and hold city representatives to account.

7.11 Conclusion: A Twin Monitoring Approach for City-Level Action and Political Engagement

This chapter has reflected upon two different methodologies for local SDG monitoring, employed by SDSN and its local US partners. The first method involved a centralized, top-down review of comparable cross-national indicators and the subsequent compilation of a US

[16] Ouyang, D. and J. Lundquist (2017) 'Data Tools for the California Bay Area', SDSN TReNDS Working Paper. SDSN: New York. Available at: http://unsdsn.org/wp-content/uploads/2017/09/180123-trends-brief-sanjose-sdg-platform.pdf

[17] As further discussed in SDSN TReNDS' related project Contracts for Data Collaboration, visit: https://www.sdsn-trends.org/blog/2019/1/22/introducing-contracts-data-collaboration

SDG Cities Index. The second method has involved local community mapping of existing indicators and metrics to the SDGs and the identification of relevant local proxies (as in San Jose, New York, Baltimore, and Los Angeles).

While the centralized, top-down approach has piqued high-level political interest, in large part thanks to media coverage, and is encouraging constructive competition among cities, the indicators used within the index are themselves too high level as to be useful for much of the day-to-day monitoring and administration of city halls. Furthermore, there are acute data gaps, partly resulting from diverse monitoring methods across cities, which local governments and stakeholders will need to fill to make the framework useful.

In San Jose, Baltimore, Los Angeles, and New York, the engagement of a broad range of city stakeholders and the consensual approach to SDG implementing and monitoring has helped to raise awareness about the SDGs and foster local buy-in and has eased integration of the SDGs into existing city plans, but it will only drive more ambitious monitoring and policy change if subsequent analysis is done to see what SDG dimensions are excluded and how they can be integrated and prioritized over time.

Lessons from both of these exercises show the utility of the SDG framework and its associated indicators for encouraging more ambitious and comprehensive sustainable development monitoring and for encouraging Mayoral engagement; however they also point to the necessity to employ a two-pronged approach to subnational monitoring of the SDGs, involving the use of headline political indicators to sustain political interest and attention, as well as more nuanced city-specific proxies to support implementation of local policies and programs.

Jessica Espey is a Senior Advisor to the United Nations Sustainable Development Solutions Network (SDSN), based in London, UK, and is the Director of SDSN TReNDS, the SDSN's expert group on data. With TReNDS, Jessica's research explores how new forms of data and technology can be harnessed to support sustainable development outcomes. TReNDS is particularly interested in building capacity in national statistical systems so they are able to capitalize on the data revolution. Jessica is a member of the Global Partnership for Sustainable Development Data's Technical Advisory Group and the Group on Earth Observation's Expert Advisory Board. Having previously worked on sustainable urban development, she has also just served as official liaison for the Scientific Steering Committee of the IPCC Conference on Cities and Climate Change. Prior to joining SDSN, Jessica served as a special adviser on the post-2015 agenda within the Office of the President of Liberia, supporting the work of the High-Level Panel of Eminent Persons (of which President Sirleaf was co-chair) and the development of the Common African Position on the Post-2015 Agenda. She has also worked as a senior researcher at Save the Children UK, the Overseas Development Institute (ODI), and the British Institute in Eastern Africa (BIEA). Jessica is a current Doctoral Candidate at the University of Bristol. She also holds a Bachelor of Arts degree with Honors in Modern History from the University of Oxford and a Master of Sciences degree in the Political Economy of Development from the School of Oriental and African Studies, University of London. She has lived and worked in Liberia, Kenya, Rwanda, the UK, and the USA.

8 Data for Good for All: Enabling All Communities to Track Progress Toward SDG Implementation

Jennifer Temmer and Stefan Jungcurt

8.1 Introduction

In this chapter, we explore how Community Indicator Systems (CISs), online platforms that communities use to share and visualize data to inform policies and decisions at the local level, can facilitate and drive localization of the Sustainable Development Goals (SDGs). Using our decade-long experience with *Peg*,[1] as a case study, we explore the conditions under which CISs can succeed in stimulating local action for SDG implementation and support measuring progress toward the SDGs. A key principle of the 2030 Agenda for Sustainable Development and the Sustainable Development Goals is that implementation should take place from the bottom-up. Within the global framework of 17 Goals and 169 targets that the international community adopted in September 2015, actors at all levels are encouraged to develop their path to making progress on global challenges. Provinces, regions, cities, communities, and other sub-national entities are encouraged to "localize" the SDGs, that is, to define local challenges and priorities within the context of the SDGs and to develop locally appropriate strategies for SDG implementation. This principle is both an opportunity and a major challenge for communities. The SDGs can act as a powerful driver of positive local change; however, unlocking this potential requires translating the SDGs into the local context and establishing monitoring systems that are meaningful to local users, while allowing reporting that contributes to assessments of progress at the international level.

8.2 A Bottom-Up World

Although a global effort, the ability for nations to tailor their approach is a trademark feature of the SDGs. Several years after the adoption of the SDGs, the bottom-up approach is starting to gain traction. Many UN member states have developed and submitted a Voluntary National Review (VNR) to the UN High-level Political Forum on Sustainable Development (HLPF).[2] VNRs are the national plans through which governments detail their national priorities, targets, strategies, and metrics to assess progress. At the sub-national level, several cities, including New York City and three Japanese cities, have developed VNRs

[1] Peg is a CIS for the City of Winnipeg, in Manitoba, Canada.

[2] The main body for reviewing progress toward the SDGs.

J. Temmer
International Institute for Sustainable Development, Simcoe, ON, Canada

S. Jungcurt (✉)
International Institute for Sustainable Development, Gatineau, QC, Canada
e-mail: sjungcurt@iisd.org

D. B. Abraham, S. D. Iyer (eds.), *Promoting the Sustainable Development Goals in North American Cities*, Sustainable Development Goals Series, https://doi.org/10.1007/978-3-030-59173-1_8

(IGES 2018). Communities, in various countries, are also localizing the SDGs through the development of community dashboards (Hawaii, USA; San Jose, USA; British Columbia, Canada) and local data hubs (New York, USA; Cambridge, Canada; and Winnipeg, Canada). These efforts are demonstrating that localizing the SDGs through measurement and reporting is a promising way to engage local actors and community members in action for SDG implementation.

While the reports of successful SDG localization are encouraging, they also reveal a fundamental challenge in measuring and reporting progress at the local level. The UN SDG indicator framework, a collection of 232 indicators that countries can use to report national progress on the 169 SDG targets, is not suitable to monitor progress in a local context. Many of the UN SDG indicators require data that is not available at the local level or is difficult to collect. Other SDG indicators may not resonate with local communities in all global regions, and risk ignoring important areas for action. Homelessness, for example, is a key issue that affects community well-being in North America, and many communities are beginning to track the number of homeless people to inform decisions on investments in housing, social support programs, and emergency shelters. Under the SDGs, however, homelessness is combined with other forms of inadequate housing into one indicator on a target referring to inadequate housing, access to services, and conditions in slums. Reductions in homelessness alone, while vital for many communities, are not well reflected in the context of the UN SDG indicator framework. The UN SDG indicator framework is a highly efficient system for measuring global progress, but it does not capture progress on the specific issues that are vital for community well-being in the eyes of community members.

How then can communities track progress on their own initiatives for SDG implementation in a way that links local conditions and priorities to the shared global aspirations and targets represented in the SDGs? One solution is to leverage existing local community-driven indicator systems that link data on local progress to national targets. Over the past decade, many communities around the world have developed Community Indicator Systems (CIS), online platforms that communities can use to share and visualize data to inform policies and decisions at the local level. CISs have evolved into an important tool for citizens to access, understand, and share information about their communities and stimulate action on key issues of community well-being. While CISs have helped facilitate positive change in many communities to date, their high cost has made them inaccessible for smaller communities or communities in poor countries.

In the following sections, we report on our experiences in "retrofitting" an existing CIS to allow users to interpret local data in the context of the global SDGs and to reduce the cost and other barriers that may prevent more widespread use of CIS to track community well-being and SDG implementation at the local level.

In 2013, the International Institute for Sustainable Development (IISD) in partnership with the United Way of Winnipeg (UWW) developed *Peg*, an online CIS for the citizens of Winnipeg, the largest city of the province of Manitoba, Canada. The data, indicators, and themes used to populate *Peg* were developed during a 2-year process of consultations and co-creation with community partners to ensure that the platform captures the issues that Winnipeggers care about most and that are important to track the city's vitality and well-being. In 2018, after winning multiple awards and gaining widespread recognition in Winnipeg and beyond, *Peg* was redesigned in a more user-friendly platform that links the existing indicator system to the SDGs. At the same time, IISD began exploring how the experience with *Peg* could be replicated in other communities in Canada and in other countries such as Peterborough Ontario, Cape Breton Nova Scotia, Trinidad and Tobago, and El Salvador.

We use *Peg* as a case study to explore how, and under what conditions, CISs can support localizing the SDGs. In this work, localizing is a process of creating linkages, through data and other information, between local concerns and priorities and national goals, so that the SDGs can assist in driving local change and serve as a framing reference against which to report and compare local progress.

This chapter begins by reviewing recent research on the role of community well-being and how measuring community well-being can support localizing the SDGs. The subsequent sections tell the story of *Peg* in three parts: (1) preliminary work and development of an indicator framework that measures what matters to Winnipeggers (2009–2012); (2) the development and use of the initial *Peg* platform (2013–2018); and (3) the redesign of *Peg*, to link the existing indicators to the SDGs, and the process of developing a software tool that facilitates the development and maintenance of other CIS portals.

We assert that, because CIS have long supported data-informed decision-making to drive community well-being, they are a well-suited tool to assist communities in tracking local SDG progress. Likewise, the SDGs, as a global movement to spur action on sustainable development concerns, provide a lens by which to reinvigorate interest in data and action on local priorities in communities around the world.

8.3 Measuring Community Well-being to Localize the SDGs

8.3.1 Community Well-being

Community well-being is a concept connected to locally shared culture, norms, and values. Indigenous communities around the world identify with "The Good Life" as an essential element of their cosmovision. Andean groups call it "Sumak Kawsay" or "Buen Vivir" perceiving well-being as a way to live in harmony with oneself, others, and with nature (Altmann 2013). In Canada, the Cree people call the good life "Mino-pimatisiwin." This concept similarly encompasses the notion of harmony, respect, growth, and healing by all. Here, relationships are a key value, underlining the importance of community (Deer and Falkenberg 2016). Western cultures tend to view community well-being as a balance between individual and community prosperity, under a backdrop of social, economic, environmental, spiritual, and cultural lenses (Lee and Kim 2015). Considering the broad nature of the definition of community well-being, there is little surprise that measurement is equally as complex.

8.3.2 Measuring Community Well-being

Community indicator frameworks have grown out of an understanding that national-level indicators of the economy or GDP alone are insufficient measures of community health and vitality (Kim 2016). Quantitative definitions of community well-being most commonly include themes of social vitality and public engagement, employment, education, housing affordability, the natural environment, physical and mental health, sport and cultural activities, time use, and basic needs (Frankish et al. 2002; CIW 2016). There are as many interpretations of community well-being as there are communities to define it. At its most basic, measuring well-being is the "process of measuring the status of the community as it pertains to the goals for the community" (Perry and Temple 2015, p. 6). Every community has unique issues, values, and goals, and thus, the indicators used to measure well-being often vary from place to place (Ibid.).

The literature points to five predominant thematic indicator measurement movements that have influenced the field of community well-being measurement. These are Quality of Life, Healthy Communities, Sustainability, Government Performance and Benchmarking and "Subjective Well-being," which incorporates Public Happiness and Life Satisfaction. CIS systems have attempted to collect data for indicators based on these themes (Warner 2014; Cummins et al. 2002). Localizing the SDGs can connect these five thematic measurement movements to the Global Goals.

8.3.3 The Sustainable Development Goals

The 2030 Agenda for Sustainable Development, which includes the Sustainable Development Goals (SDGs), was adopted by all UN member states in September 2015. The SDGs are an ambitious framework with 17 goals, 169 targets, and 232 accompanying indicators. The framework is holistic, integrated, and universal in nature. Unlike their predecessor, the Millennium Development Goals (MDGs), the SDGs address changes and challenges that all countries should address. They express aspirations for humanity as a whole and provide concrete, measurable targets, with a target date of 2030 to put the world on a pathway toward achieving these aspirations. Implementing the SDGs is thus the collective responsibility of all countries and all communities within them, irrespective of their social, economic, or environmental situations (UN-DESA 2018).

8.3.4 Localizing the SDGs

While the 2030 Agenda is a global effort, its success hinges on local commitments, investments, actions and cooperation, and engagement by actors across multiple stakeholder groups. SDG localization is key to achieving the 2030 Agenda and is particularly strategic as regional and local governments play a large role in service provision, education, health care, and ensuring a good quality of life for citizens (UCLG 2017).

SDG localization is a process whereby the SDGs are adapted, implemented, and measured at the sub-national or local level (UNDP 2018; UN-Habitat and Global Task Force 2018). The Global Taskforce of Local and Regional Governments has stressed that SDG localization is essential for achieving the 2030 Agenda and that local and regional governments can accelerate this process (UCLG 2018). A recent review of sub-national and regional governments' role in SDG implementation highlighted the role of local governments in developing pro-poor policies, raising awareness, and increasing availability of and access to local data. The review also highlighted that bottom-up approaches are more effective than implementation from the top-down (UCLG 2017). The countries who presented Voluntary National Reviews (VNRs) in 2016 and 2017 comprised approximately 400,000 sub-national governments representing over 5.2 billion citizens. These local actors play a strategic and important role in realizing the objectives of the 2030 Agenda, including monitoring indicator progress against realistic targets for each locality (UCLG 2017). At the HLPF in 2018, New York City formally presented the first Voluntary Local Review (VLR) (New York City Mayor's Office for International Affairs n.d.).

Around the world, variations of SDG localization processes are taking place. Of the 63 countries who submitted VNRs in 2016 and 2017, 38 countries reported on local government participation (UCLG 2017). On the African continent, countries such as Somalia and Tunisia are taking on SDG localization activities through research, policy and strategy development, public education, and community-based projects (SIDRA and UNDP Somalia 2018; UNDP Tunisia 2018). In Latin America, Colombia is considered a leader in monitoring regional SDG implementation through the SDG Colombia Platform (Government of Colombia 2018).

In Europe, regional programs from 15 local authorities in Germany's North Rhine-Westphalia area have developed sustainability strategies based on the SDG framework (European Union 2018).The City of Amsterdam, Netherlands has committed to SDG localization by supporting social innovators to raise awareness about the SDGs and track both existing and emerging initiatives across the city. City-led programs such as "the Action Program on Social Entrepreneurship" and "Amsterdam Impact" create opportunities for idea exchange, promotion, and access to funding (Social Challenges EU Innovation Platform n.d.).

Across North America, SDG localization is occurring at different rates using a variety of methods. Early adopters of the data dashboard process include Baltimore, New York City, San José and the State of Hawaii in the United States, and Winnipeg, Manitoba, and Kelowna, British

Columbia in Canada (Temmer 2018; Nixon and Ruckstuhl 2016; Victoria Foundation 2018; New York City Mayor's Office for International Affairs n.d.; SDSN et al. 2016; Stanford University 2017). Local authorities and civil society groups have adopted numerous methods to measure the implementation of the SDGs across the continent. According to the Taskforce of Regional and Global Governments, despite progress being made globally, local SDG initiatives are still limited. More support is needed in the form of capacity and knowledge sharing, policy guidance, and financial resources to gain momentum for the SDGs at the local level (UCLG 2018).

8.3.5 SDG Localization Methods

While there are few documented "best practices" to follow when implementing the SDGs at the community level, local governments have identified a few generic guidelines that cities can pursue. Increased participation and engagement with citizens; coordination between all levels of government; and the adoption of a rights-based approach and alternative policy development can help local and regional governments play a larger role developing more resilient and sustainable communities and can increase momentum for achieving the SDGs (UCLG 2018). SDG localization implementation toolkits refer to the need for public participation and awareness raising, development of and advocacy for a local SDG agenda, a clear implementation plan, and mechanisms for tracking progress (SDSN 2016; GTLRG et al. 2016).

Community Indicator Systems as a Tool for SDG Localization

Community indicator initiatives took root in the 1960s and 1970s and have been at the forefront of using local data to generate knowledge and action around community sustainability and well-being since that time (Wray et al. 2017). Web-based Community Indicator Systems (CISs) have developed as a logical progression in the information age. In the context of this research, CISs are defined as online platforms that curate, and make publicly available, data for indicators representing key aspects of well-being in a specific geographic location. Existing CISs are well-suited for SDG localization efforts as they provide an existing base of local, verified data that can be aligned with the SDG framework and have been adopted by local stakeholders. Likewise, adoption of a localized SDG framework can provide a new lens through which to promote local action (Iyer 2017).

By providing easy access to local-level data, CISs help improve local government transparency and accountability; they encourage public engagement, educate citizens, and inform decision-making (Holman 2009). CISs also act as a shared measurement system for collective impact efforts. "Collecting data and measuring results consistently on a short list of indicators at the community level and across all participating organizations not only ensures that all efforts remain aligned, it also enables the participants to hold each other accountable and learn from each other's successes and failures" (Kania and Kramer 2011, p. 40).

There are three primary elements needed to develop and maintain a CIS over time: access to data, a data visualization website, and public engagement and convening of stakeholders to drive action. Each of these elements requires computer hardware and software, data, human resources, technical skill, and funds to support the project (Iyer 2017; Kingsley 1999). While technology advancements such as the development of APIs can help to reduce overall maintenance costs, CIS still require significant resources to maintain the website, update data, deliver an effective communication plan, and monitor impacts.

8.4 Case Study

8.4.1 Overview

This section illustrates how a community indicator system can be used to localize the SDGs, based on the experience with *Peg* (www.mypeg.

ca). *Peg* is a community indicator system for Winnipeg, Canada, led by the International Institute for Sustainable Development (IISD) and United Way Winnipeg (UWW). *Peg*'s mission is to "track progress on key community indicators and inspire action for lasting and positive change." *Peg* was officially launched in 2013, after 2 years of community engagement to determine the indicator framework, and has acted as a sign post for measuring well-being and sustainability across the community since that time. In 2018, *Peg* was relaunched with the new Tracking-Progress CIS platform and became the first Canadian city to track local progress on the SDGs. Currently, *Peg* collects and posts data for over 60 indicators within 8 theme areas, highlights linkages between the indicators and the SDGs, and provides data that supports local decision-making and action.

8.4.2 Peg: 2009–2012 Determining What to Measure

Peg's indicator framework was developed between 2009 and 2011 through an extensive engagement process with a diverse range of stakeholder experts and community members. This process served to explore the concept of well-being and determine which measures to implement. The indicator selection process was aided by various thematic indicator working groups. Meetings were held with each group to introduce the concept of a CIS, review background research for each theme, and determine the final set of indicators. The original framework consisted of 8 theme areas with 64 indicators.

While the project team considered the indicator set to be representative of overall well-being, there were data gaps across the theme areas. These gaps existed because either the data did not exist; there was a reluctance from data holders to share information; there was concern that sharing the data may result in potential harm; the data was not collected at regular intervals; there were changes in data collection methodology; or the data required complex calculations. Over time, *Peg* has taken an incremental approach to the indicators by annually reviewing existing indicators to ensure data quality and consistency and filling potential data gaps where possible.

8.4.3 Peg 1.0 (2013–2018) Measuring What Matters

Between 2013 and 2018, *Peg* has highlighted the importance of using local data to improve community well-being in Winnipeg. Throughout *Peg*'s first iteration, maintaining technology and updating data consumed a large portion of the team's resources, placing a limited focus on communications. Despite this challenge, through regular media connections, education-based programming, and partnerships to develop annual reports, the CIS has developed a reputation as a trustworthy source for local data and a tool for decision-making.

Resources

The resources necessary to maintain *Peg* over the course of a year include staff time, financial resources, and technical expertise. The *Peg* project team consists of six core staff with varying amounts of time committed to the project. There are three project staff from each partner organization (IISD and UWW) with input and support from both organizations' leadership. Primary responsibilities include indicator updates and technology, communications, and community engagement.

Technical expertise was needed to navigate *Peg* 1.0 software for updating indicators and interpreting data trends. One challenge for the team was balancing resource allocations for data updates and technology maintenance relative to communications activities. As many resources went to data updating and system maintenance, fewer resources and staff time were available to engage with Winnipeggers around trends in the data.

Technology

The initial technology, innovative for its time, was a custom-built, ontology-based system running on a Drupal platform with indicator updates done using protégé, an open-source platform developed by Stanford University. Because of the complicated nature of the system, *Peg*'s infrastructure was stored and maintained on servers hosted by the web developer, and regular troubleshooting was needed to keep the CIS online.

Looking back, while appropriate for its time, the overall system was expensive and time-consuming to develop. The complexity of the technology increased overall annual project costs with expenses related to maintenance, troubleshooting, and technology updates.

By 2016, the technology used to run several key elements of the site became obsolete to the point that the website's front end became unusable. This required an update to the front-end design. As a result, the original indicator *wheel* (Fig. 8.1) was replaced by an updated *tile* format (Fig. 8.2) in 2017. This temporarily resolved issues posed by the obsolete Flash plugin. Further technology upgrades became necessary when the Flash-based platform used to develop the graphs was no longer accessible, thereby making data updates impossible.

Impact-Inspiring Action

Peg's tagline, "Tracking Progress, Inspiring Action," speaks to the team's ambition to implement positive change in the community through data. *Peg* aims to inspire action by informing, educating, engaging, and collaborating with organizations, decision-makers, and community members.

Peg has been used as a tool to support the work of a number of initiatives within various sectors including: community development, health, education, and government. For example, three key indicators were used to develop the case for support for the *For Every Family* initiative. This is a government and community partnership to enhance accessibility and programming at 24 family resource centers throughout Winnipeg. Secondly, the 2016 *Peg Our City* report on health equity (a collaboration between *Peg* and the Winnipeg Regional Health Authority (WRHA)) has provided the WRHA with educational materials to discuss issues of poverty, inequity, and the social determinants of health with local health staff, in order to promote better

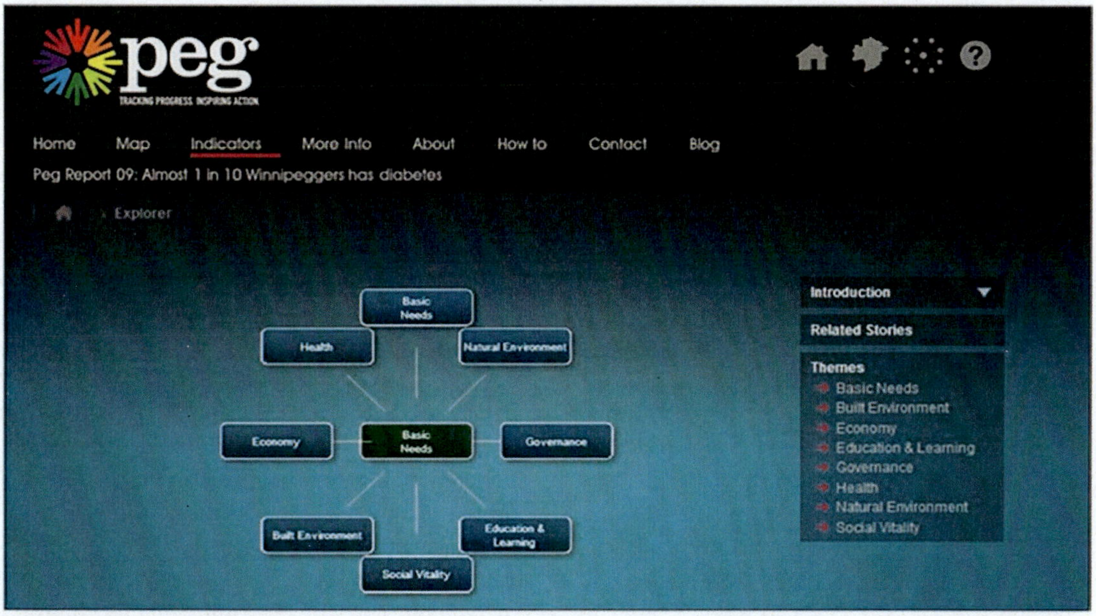

Fig. 8.1 Front end of Peg 1.0 version 1

Fig. 8.2 Front end of Peg 1.0 version 2

understanding and empathy for patients. In addition, *Peg*'s work in the Winnipeg district school board has provided social studies teachers with resource materials and tools to assist students in learning about their neighborhoods through data and carrying out local action projects in their communities. Finally, over the years, the City of Winnipeg has used *Peg*'s data as a source of information when developing plans and policies that impact local citizens, such as in the development of the current long-range official plan, *Our Winnipeg*.

8.4.4 Peg 2.0 and Tracking-Progress: 2018–Present

While *Peg* has been considered a model CIS with a well-structured indicator framework and user interface, it was recognized that improvements were needed for *Peg* to continue to be relevant and sustainable. In 2018, IISD and UWW saw the SDGs as an opportunity to both update the system and integrate SDG localization. IISD therefore developed a new CIS platform technology and began a process to include and localize the SDGs.

Technology

Building upon lessons from the original *Peg* website, the IISD technical team built the new *Tracking-Progress* CIS platform. In June 2018, *Peg* was the first CIS launched on the new *Tracking-Progress* platform.

Similar to the original *Peg* technology, *Tracking-Progress* is based on an open-source software platform – WordPress content management software. The ability to customize themes and indicators was designed specifically for *Tracking-Progress*; however, most features are standard WordPress plugins. Selecting open-source software reduced platform development costs and the time and technical skill needed to manage CIS content. The system is designed to develop a dataset template based on predetermined geographical boundaries. Once an indicator has been developed, updating data is a simple three-step process of (1) downloading the archived dataset, (2) updating and reuploading the new data points, and (3) performing quality control.

An important element of the *Tracking-Progress* platform is its network structure. All CIS sites are situated and maintained within one connected system. This allows for innovations and new features added at one site to be accessible to every CIS on the network. This means that as new features are developed, the entire network benefits.

Resources

While implementation of *Peg* on *Tracking-Progress* is relatively new, the *Peg* team has observed that the new CIS platform has helped to substantially reduce the time and technical expertise required to complete the indicator update process and the resources needed for system updates. The *Peg* team is also now able to update and make changes to the general website design – a task that previously required contracting a web designer. This simplification has enabled the team to allocate a larger portion of staff time and resources to communication and community outreach. Moreover, the Tracking-Progress tool reduces the initial cost of building new CIS portals by a factor of ten or more, making online CIS portals affordable for smaller communities.

The new *Tracking-Progress* system will also bring additional benefits and cost reductions to *Peg* over time. The platform architecture is a networked system, whereby all CIS sites in the *Tracking-Progress* network are jointly maintained with security and platform updates, and all innovations can be made available for participating sites. By servicing all the sites collectively under one umbrella, costs are shared and, thus, incrementally reduced as more sites come online.

Peg's work to localize the SDGs has also brought access to new funding sources, including securing a new funder for communications activities dedicated to SDG education and implementation.

Localizing the SDGs to Peg's Indicator Framework

As indicators are a core element of both SDGs and *Peg*, Winnipeg's SDG localization process began by comparing and aligning, or mapping together, the two indicator frameworks, thereby connecting the local indicators to the Global Goals. *Peg*'s existing indicators resulted from an extensive engagement process, so it was important to retain them. Embedding the SDGs within the *Peg* indicator framework enabled *Peg* to remain rooted in the community while helping to stimulate conversations and action in a new way.

Fifty-three of *Peg*'s 60 indicators are connected to 31 SDG targets. In some cases, the indicators were the same, while with others local interpretation was needed to account for available local data sources being used. An additional 13 *Peg* indicators are connected more broadly to the 17 SDGs. The SDGs to which there is most alignment are SDG 1 (no poverty); SDG 3 (good health and well-being); SDG 8 (decent work and economic growth); SDG 11 (sustainable cities and communities); and SDG 16 (peace, justice, and strong institutions). Gaps in alignment were present with most environmental SDGs including SDG 6 (clean water and sanitation); SDG 7 (affordable and clean energy), SDG 13 (climate action); SDG 14 (life below water); and SDG 15 (life on land). Other major gaps included SDG 5 (gender equality) and SDG 10 (reduced inequalities).

Inspiring Action on the SDGs

Since the launch of the new *Tracking-Progress* platform in June 2018, *Peg* has been at the forefront of SDG localization in North America. Outside of Winnipeg, the *Peg* team has shared their experience at various events, including a side event during the 2018 session of the UN High-Level Political Forum on Sustainable Development (HLPF) in New York and at international conferences and through webinars. In addition, the Canadian government highlighted *Peg* in its Voluntary National Review submission to the HLPF in 2018.

Locally, the *Peg* team has started an outreach and education strategy to showcase and explore how the SDGs are relevant to the local context. These linkages were highlighted in *Peg*'s 2018 *Our City* annual indicators report, which focused on the three pillars of sustainability, as well as at presentations to the Winnipeg Chamber of Commerce and local academic institutions. *Peg*'s

Our City report indicated positive progress on 11 of the 15 highlighted indicators, most notably in reduced individual and overall water consumption and reductions in waste going to the landfill. In addition, the *Peg* team has been working with the City of Winnipeg to align the SDGs and *Peg* indicators with the City's long-range official plan, *Our Winnipeg*. Inspired by this work, United Way Winnipeg has recently undergone a process to align its investments in the community with the SDGs. These discussions and activities have sparked new and exciting conversations about *Peg* and the SDGs with stakeholders and partners from all sectors, most notably the corporate and business community.

The development of the *Tracking-Progress* system and the integration of the SDGs are examples of how *Peg* is evolving to meet the needs of users and provide information to inspire action in the community. Going forward, the *Peg* team plans to continue its work to enhance the system and localize the SDGs. These activities will include an engagement process to review and revise *Peg*'s existing indicator set and implementation of a new, multi-stakeholder communications strategy that will showcase the alignment between *Peg* and the SDGs.

8.5 Conclusions

For over three decades, CISs have measured, informed, and inspired citizens to take action on local issues. However, the time and resources traditionally needed to maintain a CIS have limited their use in urban centers. The global push to achieve the SDGs, combined with the development of new, user-friendly technologies, the proliferation of social media, and an increased understanding of the power of data, has meant that data platforms are gaining popularity around the globe. CISs are a particularly effective type of data platform as they purposefully engage communities in developing indicator frameworks, thereby building local ownership over the community's well-being. When this bottom-up approach to community well-being measurement is paired with the SDG framework, CISs become a powerful tool for stimulating local SDG action. A CIS is most effective for tracking progress and inspiring local action when the system is designed to minimize technology and staffing costs, and it takes into consideration how local stakeholders interpret, use, and share the data. Tools such as IISD's *Tracking-Progress* platform make it easier for communities of all sizes to harness the power of data to encourage public participation and understanding of local issues.

Reflecting on *Peg*'s experience, we see that the introduction of an easy-to-use, low-cost technology for the CIS itself has been a positive element for the project's evolution. Another key element for success has been the involvement of community organizations in a position to undertake meaningful communications and outreach efforts. IISD has developed the *Tracking-Progress* platform to make CIS more widely accessible and reduce cost and efforts required to a level, allowing CIS platforms to be sustainable in the long term. The *Peg* example demonstrates why communities should invest in the technical and human resources needed to ensure that a CIS system can deliver the full benefits possible from data-driven decision-making and community ownership of efforts to track local well-being.

References

Altmann P (2013) Good life as a social movement proposal for natural resources use: the indigenous movement in Ecuador. J Sustain Dev 10(1):59–71. Retrieved from: https://www.researchgate.net/publication/237064750_Good_Life_As_a_Social_Movement_Proposal_for_Natural_Resource_Use_The_Indigenous_Movement_in_Ecuador

Canadian Index of Well-being (CIW) (2016) How are Canadians really doing? The 2016 CIW National report. Canadian Index of Well-being and University of Waterloo, Waterloo. Retrieved from: https://uwaterloo.ca/canadian-index-well-being/sites/ca.canadian-index-well-being/files/uploads/files/c011676-nationalreport-ciw_final-s_0.pdf

Cummins RA, Eckersley R, Pallant J, Van Vugt J, Misajon R (2002) Developing a national index of subjective wellbeing: the Australian unity wellbeing index. Soc Indic Res 64:159–190. Retrieved from: http://www.uvm.edu/pdodds/files/papers/others/2003/cummins2003a.pdf

Deer F, Falkenberg T (2016) Indigenous perspectives on education for well-being in Canada. University of Manitoba. Retrieved from: https://www.eswb-press.org/uploads/1/2/8/9/12899389/indigeneous_perspectives_2016.pdf

European Union (2018) Localizing the SDGs-Developing sustainability strategies at the local level. Representation of North Rhine-Westphalia to the EU. June 26, 2018. Retrieved from: https://cor.europa.eu/hu/events/Pages/localizing-SDGs.aspx

Frankish, J., Kwan, B., Flores, J. (2002). Assessing the health of communities-indicator projects and their impacts. Canadian Institute for Health Information. Ottawa.: Retrieved from CIHI website: https://www.cihi.ca/en/cphi_assessing_health_en.pdf

Global Taskforce of Local and Regional Governments, UNDP, UN Habitat (2016) Roadmap for localizing the SDGs: implementation and monitoring at subnational level. Retrieved from: https://www.global-taskforce.org/sites/default/files/2017-06/bfe783_434174b8f268 40149c1ed37d8febba6e.pdf

Government of Colombia (2018) Como va Colombia en los ODS? Retrieved from: https://www.ods.gov.co/

Holman N (2009) Incorporating local sustainability indicators into structures of local governance: a review of the literature. IJJS 14(4)

Institute for Global Environmental Studies (IGES) (2018) VLR Lab. Retrieved from: https://www.iges.or.jp/en/sdgs/vlr/index.html

Iyer SD (2017) Building a local SDG data monitoring system for Baltimore: insights from the US National Reporting Platform and New York City. SDSN. Retrieved from: http://unsdsn.org/wp-content/uploads/2017/12/180123-trends-brief-baltimore-data-monitoring.pdf

Kania J, Kramer M (2011) Collective impact. Stanford social innovation review. Leland Stanford Jr. University. Retrieved from: https://ssir.org/images/articles/2011_WI_Feature_Kania.pdf

Kim Y (2016) Community well-being Indicators and the history of "Beyond GDP". Research & Policy Brief Series Issue 73 June 2016. Cornell University. Retrieved from: https://cardi.cals.cornell.edu/sites/cardi.cals.cornell.edu/files/shared/PolicyBriefJune-16_draft5.pdf

Kingsley T (1999) Building and operating neighborhood Indicator systems: a guidebook. The Urban Institute. Retrieved from: http://webarchive.urban.org/UploadedPDF/409297-Building-and-Operating-Neighborhood-Indicator-Systems-Guidebook.pdf

Lee SJ, Kim Y (2015) Searching for the meaning of community well-being. In: Lee SJ, Kim Y, Phillips R (eds) Community well-being and community development concepts and applications. Springer, Cham

New York City Mayor's Office for International Affairs (n.d.) NYC is first city in the world to report to UN on Local steps toward global goals. Retrieved from: https://www1.nyc.gov/site/international/programs/voluntary-local-review.page

Nixon H, Ruckstuhl S (2016) San José implementing the UN's Sustainable Development Goals at the local level. SDSN & San José State University. Retrieved from: http://www.sjsu.edu/sustainable-cities/docs/sci-reports-sanjose-final.pdf

Perry J, Temple Z (2015) Developing a framework for measuring community well-being in Central Massachusetts. Worcester Community Project Center. Retrieved from: http://www.cmrpc.org/sites/default/files/Documents/CMRPC_Indicators_IQP_B15_Report.pdf

SDSN, BNIA, University of Baltimore (2016) Baltimore's sustainable future: localizing the UN Sustainable Development Goals, Strategies and Indicators. Retrieved from: https://www.dropbox.com/s/psr4n-7crbcv0tv6/170125_SCI_Baltimore_Report_Interim.pdf?dl=0

Social Challenges EU Innovation Platform (n.d.) Localise the Sustainable Development Goals (SDGs) to action. Retrieved from: https://www.socialchallenges.eu/en-US/city/19/Challenges/1077

Somalia Institute for Development and Research Analysis, UNDP Somalia (2018) Localizing the Sustainable Development Goals (SDGs) & SDG7 in Puntland. Puntland, Somalia

Stanford University (2017) SDG dashboard for San José. Stanford Sustainable Urban Systems Initiative. Retrieved from: http://sdg-stanford.opendata.arcgis.com/

Sustainable Development Solutions Network (SDSN) (2016) Getting started with the SDGs in cities: a guide for stakeholders. Retrieved from: http://unsdsn.org/wp-content/uploads/2016/07/9.1.8.-Cities-SDG-Guide.pdf

Temmer J (2018) Four ways cities are localizing the SDGs. International Institute for Sustainable Development, Winnipeg. Retrieved from: https://www.iisd.org/blog/four-ways-cities-are-localizing-sdgs

UCLG (2017) Local and regional governments' report to the 2017 HLPF: national and sub-national governments on the way towards the localization of the SDGs. European Commission. Retrieved from: https://www.uclg.org/sites/default/files/localgov_report_localizationsdg_hlpf.pdf

UNDP (2018) Human Development Index (HDI). Retrieved from: http://hdr.undp.org/en/content/human-development-index-hdi

UNDP Tunisia (2018) Strengthening and dissemination of strategic urban planning initiatives in Tunisia Madinatouna Project. Retrieved from: http://www.tn.undp.org/content/tunisia/fr/home/projects/renforcement-et-diffusion-des-initiatives-de-planification-urbai1.html

UNDP, UN-Habitat, Global Task Force (2018) What is localization? Retrieved from: http://localizingthesdgs.org/about-us.php

United Cities and Local Governments (UCLG) (2018) Towards the localization of the SDGs: local and regional governments report to the 2018 HLPF 2nd report. Retrieved from: https://www.gold.uclg.org/sites/default/files/Towards_the_Localization_of_the_SDGs.pdf

United Nations Department of Economic and Social Affairs (UN DESA) (2018) Sustainable Development Goals knowledge platform. Retrieved from: https://sustainabledevelopment.un.org/

Victoria Foundation (2018) Victoria's Vital Signs: The Future is in our hands. Victoria Foundation. Retrieved from: https://victoriafoundation.bc.ca/wp-content/uploads/2018/10/Vital-Signs-2018_LR.pdf

Warner B (2014) The future of community indicator systems. Jacksonville Community Council Inc.. Retrieved from: http://www.whatcountsforamerica.org/wp-content/uploads/2014/11/Warner.pdf

Wray L, Stevens C, Holden M (2017) The history, status and future of the community indicators movement. In: Holden M, Phillips R, Stevens C (eds) Community quality-of-life indicators: best cases VII. Springer, Cham

Jennifer Temmer is an Associate with the International Institute for Sustainable Development (IISD). Jennifer has worked with communities across North America to measure sustainability and community well-being, including the Peg Community Indicator System (CIS) and the IISD's Tracking-Progress CIS tool. She has also worked with diverse communities to improve food security through community initiatives such as youth-led gardening programs and women's micro-credit support. Jennifer is currently a Ph.D. student, researching sustainable food systems in Northern Canada at the Wilfrid Laurier University in Waterloo, Ontario.

Stefan Jungcurt leads the work of Institute for Sustainable Development (IISD) on indicators and data for SDG monitoring, including Community Indicator Systems and the Tracking-Progress platform. Since 2003, Stefan has worked for IISD in various positions, including writer and team leader for the *Earth Negotiations Bulletin*, content editor for the *SDG Knowledge Hub*, and project officer on climate change and energy. He was also a program manager and research associate at the Council of Canadian Academies. Stefan completed his Ph.D. on institutional interplay in global environmental governance at Humboldt University, Berlin, in 2007. Originally from Germany, Stefan is based in Gatineau, Québec.

9 Helping the Neighborhood: Creating a Sustainability Indicator of Substandard Housing

Luis Estevez

9.1 Introduction

Quality of housing is paramount to quality of life and to any community sustainability effort. Countries and international organizations recognize this reality in their visions and goals. For instance, in 2015, the United Nations reaffirmed the importance of addressing the issue of inadequate housing by incorporating this indicator in goal 11 of the Sustainable Development Goals (SDGs).

The importance of this study is in proposing a methodology that could be replicated by other communities to create an indicator for the purpose of implementing goal 11 of the SDGs and contributing to the literature on defining and measuring inadequate housing.

The project also reflects the collaborative efforts of different community stakeholders. This research has a direct impact on practice, as well. It demonstrates that by taking a participatory research approach, it is possible to generate local data on substandard housing in an efficient way to facilitate planning for sustainability in general and for housing in particular.

It is presumed that because of its economic prosperity compared to other countries, the presence of inadequate housing in the United States will be minimal. Progress has undeniably been made since the establishment of this national goal in the Housing Act of 1949: "A decent home and a suitable living environment for every American family." Nonetheless governmental reports, such as the *Worst Case Housing Needs* report to Congress (Watson et al. 2017), show that not only is there a reduced supply of affordable housing but this number gets reduced further due to inadequate conditions.[1] At the same time, current data and conceptualizations on which these reports are based do not represent accurate numbers because of the obsolescence of the variables used to define inadequate housing (Eggers and Moumen 2013; Emrath and Taylor 2012).

The social and economic consequences of the deterioration of housing stock in inner-city neighborhoods in cities across the United States are well known. Although causes for this deterioration are multiple, it is certain that the presence of substandard or deteriorated[2] housing can lead to blight conditions, which in turn decreases property values, affects the overall health of local housing markets, increases safety hazards, and reduces local tax revenue. Various creative strategies have been undertaken to reverse this deterioration. However, before the implementation of any strategy or policy, conducting an analysis of the quality of housing stock to identify and quantify inadequate housing is a necessary first step.

L. Estevez (✉)
St. Cloud State University, St. Cloud, MN, USA
e-mail: lestevez@stcloudstate.edu

[1] In 2017, there were only 89 adequate units available for every 100 low-income renters.

[2] The terms substandard, inadequate, or deteriorated are used interchangeably in this paper.

D. B. Abraham, S. D. Iyer (eds.), *Promoting the Sustainable Development Goals in North American Cities*, Sustainable Development Goals Series, https://doi.org/10.1007/978-3-030-59173-1_9

Quality of housing is conceived in the literature as a multidimensional concept; therefore, every method and tool to assess it depends on the researcher's specific perspective and conceptualization (Sinha et al. 2017). To consider all these dimensions in an assessment is complex, and although some good strategies have been proposed, they require an intensive collection of data (Kain and Quigley 1970).

Collecting data regarding the amount of substandard housing in a community is crucial in order not only to understand the proportion of people living in such conditions but also to have the necessary indicators in planning for sustainability. Traditional ways to identify inadequate housing, such as survey fieldwork, have proved to be time-consuming and expensive (Smith et al. 2003), and although some efforts have been made to come up with a more efficient way to collect data for assessment purposes, community organizations, small ones in particular, still struggle to handle these approaches because of the methodological complexity or scale of analysis (Koebel 1986; Sumka 1977). Another problem facing communities, and specifically neighborhood associations, is the lack of resources to be rigorous when gathering data to measure inadequate housing. These types of organizations, unless they collaborate with other entities with resources and knowledge, cannot make progress in their planning for sustainability goals.

This paper presents the results and analysis of a community-based research project in a traditional inner-city neighborhood in the Midwestern United States. The neighborhood found itself in need of a rigorous assessment of inadequate housing due to its repercussions on the overall sustainability of the area. As a response to this problem, an academic institution, the neighborhood association, and local government officials collaborated in an action-based research project to provide data for planning for sustainability in an efficient and affordable way.

Using data collected through an instrument designed for this project, two composite indicators[3] were created that sought to answer two basic questions: (1) At what point does housing qualify as inadequate? (2) How many housing units in need of rehabilitation are currently in existence? The analysis and results show that both indices give similar results. Both indices provided answers to these questions, and one of them gives detailed information about the conditions of units, which in turn helps with planning for housing and policy decision-making.

Statistical tests of the relationship between the two indices confirm the validity of the indices' results. The results of the composite indicators are also compared and validated using housing inspections for code violations. The findings are relevant because of the direct impact they have on practice by enabling communities to use existing information to identify deteriorated housing. Identifying inadequate housing will provide the necessary data for decision-making in the pursuit of funding and policymaking to help neighborhoods reverse the consequences of deteriorated housing.

9.2 Relevant Literature on Indicators for Measuring Inadequate Housing in the United States

Housing is paramount to the SDG agenda. According to the 2017 report of the Secretary-General of the United Nations, "Progress towards the Sustainable Development Goals" is calculated that 1.6 billion people live in inadequate housing. Of this number, 880 million urban residents live in slum conditions. Therefore, the New Urban Agenda places a priority on housing concerns, along with other issues such as poverty.

Housing censuses are the primary sources of disaggregated data to measure and track indicators of inadequate housing. According to the same United Nations report mentioned above, during the 10-year period from 2007 to 2016, 89% of countries around the world conducted one housing census, while 25 countries did not have such a data source.

[3] The terms composite indicator and index are used interchangeably in this paper.

Although national censuses are important sources of data in indicators for the SDGs, it is at the local or community level where the implementation of SDGs could have the most impact. It is important to keep in mind the issues around the adoption and implementation of SDGs at the community level. Local jurisdictions in the United States are notable for enjoying great autonomy in decision-making, adoption, and implementation of policies, as long as these do not contradict state or federal laws or statutes. This autonomy becomes an issue in the adoption of SDGs. Unless SDGs are part of the vision of a community, the implementation of such goals at the community level will be difficult. In addition to these issues of governance and implementation, as this section will show, conceptualization of inadequate housing is fundamental in any effort to collect data in such indicators.

Currently, the only source of data regarding quality of housing in the United States that is considered for SDG purposes is that collected by the American Housing Survey. Nonetheless, as this section will also show, the data is limited to specific geographic areas and has been recently criticized for not properly measuring inadequate housing and therefore undercounting the presence of inadequate housing nationwide.

Measurements of inadequate housing are constrained both by their conceptualization and the availability of indicators on housing quality. The literature shows three general approaches that studies have used to provide indicators of quality of housing: (1) studies that traditionally relied on the use of official data from the US Census first and the American Housing Survey later, (2) studies that have undertaken their own housing quality assessments, and (3) studies that have come up with strategies by using alternative or local indicators related to housing quality. These approaches are not mutually exclusive; on the contrary, some studies have combined indicators from two or three different sources. In general, the literature attests to the changing nature of the concept and how this in turn obscures the apparent progress in addressing inadequate housing in the United States.

Following is a brief review of the most notable cases using the previously mentioned approaches or a combination of them. Special attention is given to the most recent literature with a focus on the use of alternative indicators or secondary data, specifically, the use of indicators contained in administrative records of tax assessments. These last strategies have emerged as a response to the time and cost limitations of performing traditional housing assessments and the lack of official data for some areas, especially at the neighborhood level.

Although early efforts to collect data to identify inadequate housing in the United States can be traced to 1892, it was not until 1936 when the first nationwide effort took place with a survey of housing conditions as a Works Progress Administration (WPA) project. This survey served as a prototype for the first census of housing in 1940 (Simonson 1981). Between then and the 1960s, the US Census collected information on housing quality. As a result of some inconsistencies in the way data was collected, the Bureau dropped[4] its indicators of housing quality for the 1970 Census (Sutermeister 1969).

The fluid and problematic nature of what is considered inadequate housing has been evident since early studies examining the data collected in quality of housing surveys. For instance, in a report presented to the National Commission on Urban Problems, Kristof (1968) questioned the validity of the current criteria of housing needs, in particular the concept of substandard housing as defined by the US Census Bureau. Similarly, Weicher (1980), when verifying progress in addressing inadequate housing between 1940 and 1970 using the census data, also observed that this apparent improvement is obscured by the problem of defining and measuring the overall physical condition of the units.

Based on this experience with the US Census, by 1973, the Department of Housing and Urban Development (HUD) and the US Census Bureau joined efforts and started conducting the American Housing Survey (AHS), which, among other indicators, collects data on about 30 differ-

[4] Although some data still was collected.

ent kinds of housing deficiencies. Unfortunately, AHS data is not useful for local communities or neighborhoods because it is limited to specific metropolitan areas.

From the beginning of its implementation, scholars and analysts began to use the AHS information in different ways and for different purposes. Early analysts began to measure inadequate housing by creating their own AHS-based definitions (see Levine 1978). Simonson (1981) records a detailed narrative of early studies and their definitions using indicators from the AHS. What is relevant in these early efforts using AHS data is that depending on the indicators used to conceptualize inadequate housing, the number of units could increase or decrease, again proving the effect of the changing nature of the concept.

In addition to the efforts to define and measure inadequate housing using AHS indicators, other scholars began to use the data to assess the statistical relevance of individual indicators. For instance, Newman and Struyk (1983), in their study addressing the relationship between poverty and housing deprivation, relied on AHS's definition. Newman and Schnare (1988), also using AHS information and definitions, examined the relationship between income and housing assistance programs. As an example of the combination of sources of indicators, Bianchi et al. (1982) examined the degree to which racial differences in housing narrowed over time using 1960 Census information and AHS data.

Scholars have also used AHS data to test the validity of alternative measures of inadequate housing (e.g., housing inspection assessments) by comparing them against the classification used by HUD. For instance, using information on the physical conditions of housing units in the Experimental Housing Allowance Program (EHAP), Budding (1980) found a higher number of inadequate housing units compared to estimates using AHS data.

Similar to Weicher's (1979) findings about the problem of definition and the apparent decrease in inadequate housing, using US Census housing information, two more recent studies found inconsistencies in the accuracy of the indicators used to identify inadequate housing as conceptualized by the AHS. In a work prepared for HUD, Eggers and Moumen (2013) found a decline in the presence of inadequate housing after the definition was instituted by the AHS. The authors also express a concern for how this apparent improvement has decreased interest in the topic. This lack of urgency has been reinforced by the apparent decline in the proportion of severely inadequate housing among low-income households (Orr and Peach 1999).

Empirical work by Emrath and Taylor (2012) found not only that the AHS's indicators of physically inadequate housing units are not significant but they also had the wrong signs. The authors propose instead new measurements of housing inadequacy by using other indicators contained in the AHS. The proposed new definition reveals a greater number of housing units in the United States as being physically inadequate. Emrath and Taylor's new approach of using only indicators of external physical conditions is the approach taken in this study for the creation of the housing quality composite indicator.

Scholarly work using AHS data is abundant. But although the AHS information is exhaustive and contains data for a good period time regarding the same housing units, this is limited to specific metropolitan areas, making it not useful for analysis of other geographical areas. With this limitation from the AHS in mind, other scholars began to rely on other data sources, specifically on the implementation of housing assessment surveys. A notable study of this type is the work by Kain and Quigley (1970), who relied on 39 variables provided by three separate surveys for the City of St. Louis. The concerns with these types of surveys or assessments lie both in terms of time and cost of implementation and the potential errors made by collectors in the field.

Considering the geographical limitations of official information and the disadvantages of lengthy surveys, scholars have attempted to come up with different strategies using alternative indicators as a more efficient way to identify inadequate housing. Relying on the latest empirical work on determinants of quality of housing, scholars have made use of property tax information for measuring inadequate housing (Koebel 1986; Smith et al. 2003; Sumka 1977; Zwick and Schneider 1990; Kutty 1999). The works by Koebel (1986) and Smith (2003) are of particular interest for this study because they are the most promising for implementation at the neighborhood level.

Koebel's work (1986) took a hedonic modeling approach to create an index of housing quality using 22 variables combining property taxes and census information. He later validated the index by comparing it with a record of housing inspections for code violations. In his work, Koebel found the use of this strategy at the metropolitan level problematic because of the variation of housing stock. Unfortunately, Koebel's model misclassified some units and is not very easy to implement by small organizations that do not necessarily have the analytical skills to build such an index. Koebel's strategy of using code violations to verify the results of his index is used in this case study.

Smith et al. (2003) performed a descriptive analysis using the American Housing Survey (AHS) information and tax assessor records for the Tampa-St. Petersburg metropolitan area. Using the indicators suggested in recent literature as the most relevant (area of unit, cost of new construction, and land and housing values) the authors propose a measure – the ratio of market value to unit value new – which then they used to estimate the proportion of units considered inadequate. The authors contrast their results against the data reported by the AHS. Although in principle the authors aim to create a single measure to estimate the quantity and degree of substandard housing in a community at the neighborhood level, the research is mainly framed at the metropolitan level. The proposed value ratio measure is used for the creation of the second composite indicator in this paper. This case study will contribute to the limited literature using secondary indicators to measure inadequate housing at the neighborhood level.

9.3 Methodology and Data

The data for this paper comes from a community-based research project originated to support the efforts of a neighborhood association in the downtown area of the City of St. Cloud in Minnesota interested in determining the conditions of housing units in the area.[5] The neighborhood is also home to the main campus of St. Cloud State University (SCSU) (see Fig. 9.1 for location).

The neighborhood association, known as Neighborhood University Community Council (NUCC), has been concerned with the deterioration of housing, but it lacks any precise data about the amount, conditions, and the extent of the presence of inadequate housing. It is in this context that a research project emerged as a collaboration between local officials, the neighborhood association, and the SCSU. The questions driving this research effort were: When do housing units start to deteriorate, and what is the proportion of housing units in need of rehabilitation in the neighborhood? More precisely, how can one obtain an indicator of inadequate housing to support a policymaking process?

To answer these questions, the main objective driving this research project was to come up with an efficient way to create indicators that could yield detailed information and that could be easily replicated. Based on this premise and the experience in the literature about measuring inadequate housing, the strategy was the creation of two composite indicators or indices. One index was created using existing information contained in tax appraisal records, and the other index was created based on data from a traditional visual assessment of exterior conditions of units in the neighborhood.

The first composite indicator (hereafter "value ratio index") gives a fast look at the amount and age-related conditions of housing stock. The

[5]The dataset used in this paper is available at Estevez 2019.

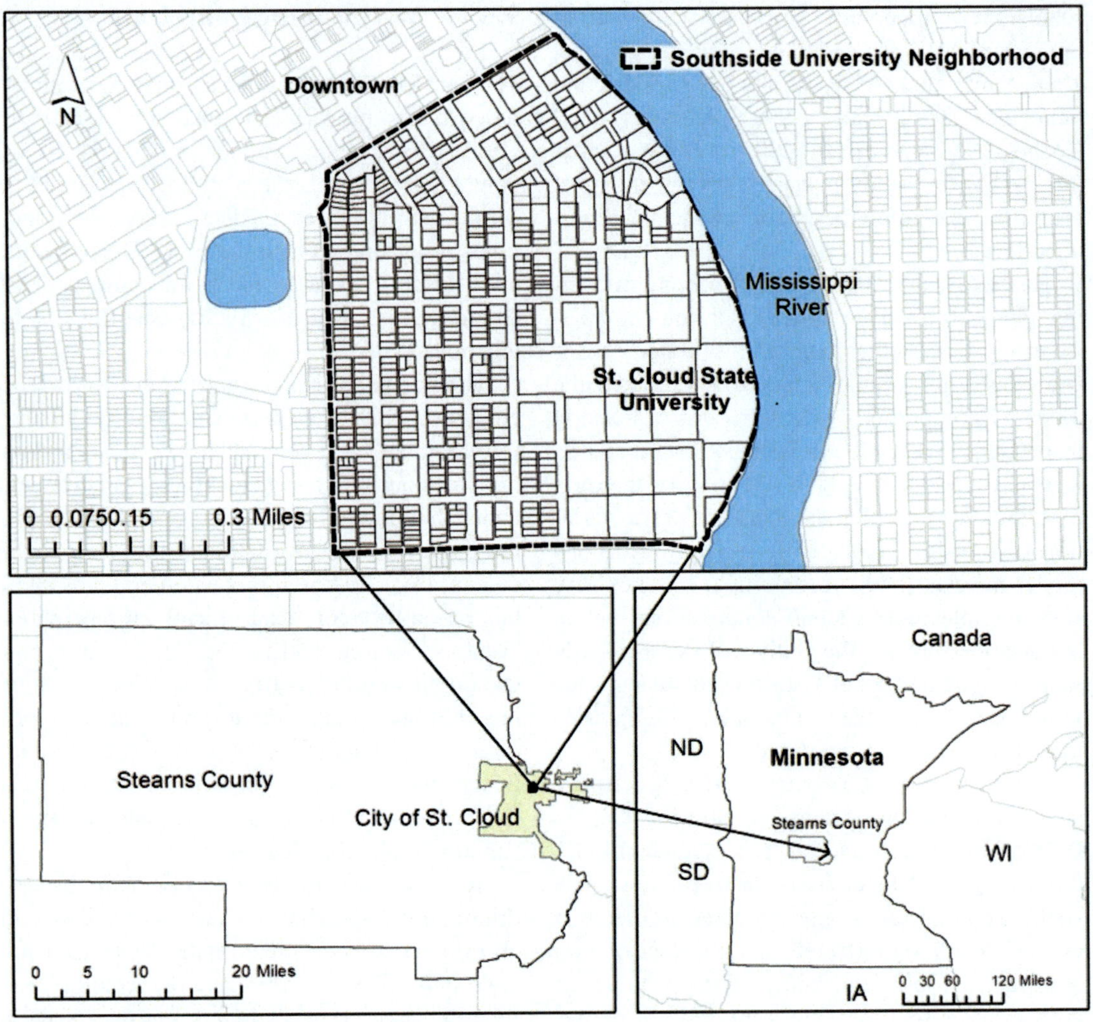

Fig. 9.1 Location – case study area

index was tested by Smith (2003) with promising results. The second composite measure (hereafter "housing quality index") gives the community not only detailed information on inadequate housing but also details about the conditions of particular dimensions. This will help the neighborhood in its decision-making process moving forward with specific goals and objectives for the area.

Recent empirical evidence in the literature shows that external physical conditions are among the most significant dimensions when determining quality of housing; therefore, the instrument was conceived under this assumption. The survey was designed in collaboration with the neighborhood association through focus groups. Elected officials offered input for the instrument as well. Once the instrument was created, it was tested in the field and adjusted to correct for mistakes and better use. The survey was implemented by students in university courses from the geography and planning department. The data was then processed and reviewed using the software IBM SPSS Version 25.

Information was collected on 362 dwelling units, from which 61 (16%) are multifamily units,[6] 149 (41%) single-family residential units, and 152 (42%) duplexes and triplexes. In terms

[6]Four or more units.

of occupancy, 70 (19%) are considered owner-occupied units and 292 (81%) rental units.

Data for the value ratio index was obtained from the St. Cloud City Assessor's Office. The database contains all the necessary property information for the creation of the index. An external variable for this index was the average price per square foot for new construction. In this case, the number used was that provided by the US Census report: *Highlights of Annual 2017 Characteristics of New Housing* (2017).

9.3.1 Composite Indicators

Value Ratio Index

Following Smith's (2003) proposal, the value ratio index can be summarized as:

Value ratio index = Current unit value/unit value new

where:

Current unit value = Assessed value of the building

Unit value new = Construction cost (per sq. ft.) × building area of unit (sq. ft.)

Housing Quality Index

The housing quality composite indicator was created by using eight out of the ten physical exterior dimensions assessed with the survey. The other two dimensions were not present in most of the assessed housing units (chimneys and detached garages). Considering these dimensions would mean that the index would be calculated for only half of the housing units in the area, limiting the analysis.[7] Figure 9.2 shows a summary of the dimensions and the proportion of housing units considered adequate and inadequate.[8] The index was created using a simple linear aggregation approach. For each adequate dimension, every dwelling unit is scored with 1 and the dimensions with inadequate a 0:

Housing quality index = D1 + D2 + D3 + D4 + D5 + D6 + D7 + D8

Table 9.1 summarizes the proportion of units according to the index scores. The interpretation is straightforward, 0 meaning the unit does not have any dimensions that have an adequate score (passing rate) and 8 meaning that the unit has an adequate score in all eight dimensions.

A fundamental assumption in the creation of the housing quality index was the specification of when a unit starts to be considered overall inadequate. For this purpose, the number of housing units obtained by the value ratio index was considered as a threshold. According to Smith (2003), units in need of rehabilitation started to appear when the index fall below 0.45. Following this assumption, different scenarios were run together with the value ratio indicator to see at what specific point units for the housing quality index could be considered inadequate. The threshold in this case was all units with an index score of 6 and lower (this means 6 or less dimensions with an adequate score). This threshold happens to be in line with what the neighborhood considered a reasonable limit to define adequate and inadequate housing.

9.4 Results and Analysis

Because of the interest in knowing not only the number of housing units in need of rehabilitation but also how these are distributed according to type and occupancy, the results presented considered this disaggregation in its analysis. For the housing quality index, only 345 out of 364 dwellings units had a score in all eight dimensions; therefore, an index score was calculated for all of them. In terms of occupancy, 225 (76%) are rental properties and only 70 (24%) owner-occupied.[9] Regarding the type of housing units,

[7]Imputation is an alternative in the absence of missing indicators. This methodological procedure was not considered due to the aim of creating an index that could be easily replicated by the organization.

[8]The survey contained a Likert scale for the ten dimensions. The values were 1, deteriorated; 2, poor; 3, average; 4, good; and 5, excellent conditions. For the purposes of creating a simple index, these categories were transformed into only two: inadequate and adequate. 1 and 2 scores inadequate and 3, 4, and 5 adequate.

[9]An assumption to classify owner-occupied units was to define in this way those cases claiming homestead tax

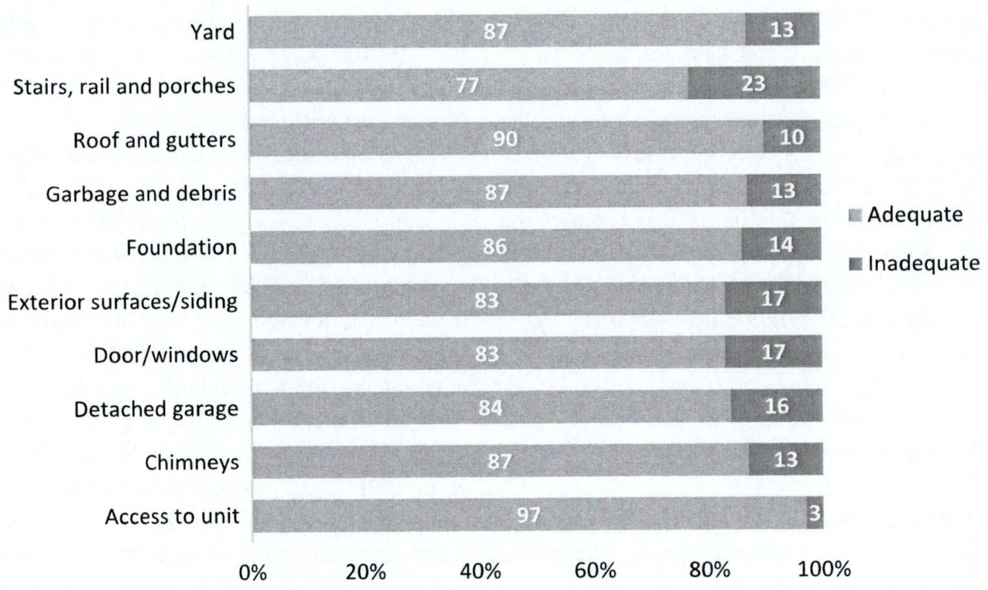

Fig. 9.2 Proportion of housing units and the ten dimensions

Table 9.1 Housing quality index scores and proportion of housing units

Index score	Units	Percent	Cumulative percent
1	5	1	1
2	10	3	4
3	8	2	6
4	13	4	10
5	27	7	17
6	39	11	28
7	64	18	46
8	198	54	100
Total	364	100.0	

147 (43%) are single-family residential, 148 (43%) duplexes and triplexes, and 50 (14%) multifamily units.

Regarding the value ratio index, 362 housing units were considered. By occupancy, 231 (77%) are rental and 70 (23%) owner-occupied. For type of housing units, 149 (41%) are single-family residential, 152 (42%) duplexes and triplexes, and 61 (17%) multifamily.

Another important element to consider in the analysis was age of buildings. Therefore, buildings were grouped by age following AHS and Smith's (2003) experience when analyzing when inadequate units start to appear in the housing stock. Units were aggregated according to their age in groups. The first ten groups (groups 1 through 10) are for every year a unit was built. Groups 11 through 15 consider 5-year cohorts. Groups 16 through 25 are for units built every 10 years. The oldest building in the data is 151 years old. The information about age reveals the context of the neighborhood, a Midwestern old downtown area (see Fig. 9.3 for age groups). It is important to note that no unit appears in groups 1–11 because the most recent unit was built 17 years ago.

The benefit of considering age is twofold. It provides the amount of inadequate housing units at a specific point in time, and it allowed for later statistically testing the difference between units at different ages.

Both composite indicators yield an almost identical number of units in need of rehabilitation (value ratio index = 30% and housing quality index = 29%). Both indices detected an increase in the presence of inadequate units as the stock grows older. However, while the value ratio index found an increasing proportion of units in need of rehab the older the stock got all the way to the

exemption in the administrative records of tax assessments.

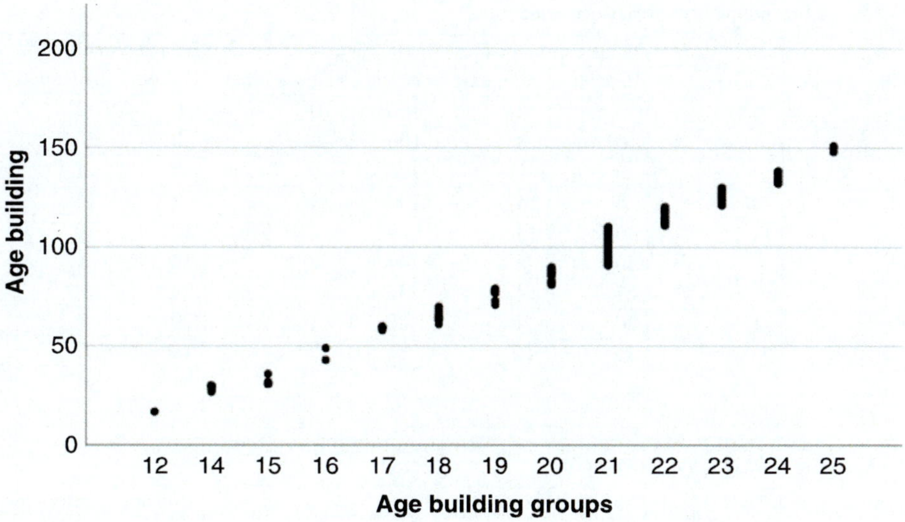

Fig. 9.3 Building age group plot

Table 9.2 Age group and quality: owner-occupied and rental units

	Value ratio index				Housing quality index			
Age building groups	Needs rehab	Percent	Adequate	Percent	Needs rehab	Percent	Adequate	Percent
15	1	33	2	67	0	0	1	100
17	0	0	5	100	0	0	5	100
18	0	0	11	100	2	18	9	82
19	0	0	10	100	1	10	9	90
20	4	12	29	88	6	18	27	82
21	25	23	84	77	38	35	70	65
22	22	49	23	51	17	39	26	60
23	26	50	26	50	17	33	34	67
24	13	43	17	57	5	17	25	83
25	1	33	2	67	0	0	3	100
Total	92	30	209	70	86	29	209	71

age group 24, the housing quality index had a decrease in age group 24.

It is important to point out the inconveniences of the lower number of units at some age levels. For instance, age groups 15 and 25 had less than five units to be observed. These were the groups with the greatest difference between the two indices (Table 9.2).

In terms of the results for both separate owner-occupied and rental units, Tables 9.3 and 9.4 show all the values, taking into account occupancy. As regards owner-occupied units, the results from the value ratio index and the housing quality index are different, 36% and 19%, respectively. In the case of rental units, the results are almost identical (value ratio index = 29% and housing quality index = 32%). Again, some age groups present less than five housing units (age groups 17, 18, and 19 in the case of owner-occupied and 15, 17, and 25 in the case of rental).

Tables 9.5, 9.6, and 9.7 present the number and corresponding proportions of units that are adequate or in need of rehabilitation by type of housing units. Regarding single-family residential units, both indices detected the same proportion of units in need of rehabilitation (26%). Only age groups 17 and 25 had less than 5 units total.

When compared to the proportions for single-family residential units, for duplexes and triplexes, the total number of units in need of

Table 9.3 Age group and quality: owner-occupied units

		Value ratio index				Housing quality index		
Age building groups	Needs rehab	Percent	Adequate	Percent	Needs rehab	Percent	Adequate	Percent
17	0	0	2	100	0	0	2	100
18	0	0	3	100	0	0	3	100
19	0	0	1	100	0	0	1	100
20	0	0	14	100	1	7	13	93
21	11	46	13	54	4	17	20	83
22	5	56	4	44	3	33	6	67
23	8	67	4	33	4	33	8	67
24	1	20	4	80	1	20	4	80
Total	25	36	45	64	13	19	57	81

Table 9.4 Age group and quality: rental units

		Value ratio index				Housing quality index		
Age building groups	Needs rehab	Percent	Adequate	Percent	Needs rehab	Percent	Adequate	Percent
15	1	33	2	67	0	0	1	100
17	0	0	3	100	0	0	3	100
18	0	0	8	100	2	25	6	75
19	0	0	9	100	1	11	8	89
20	4	21	15	79	5	26	14	74
21	14	16	71	83	34	40	50	59
22	17	47	19	53	14	41	20	59
23	18	45	22	55	13	33	26	67
24	12	48	13	52	4	16	21	84
25	1	33	2	67	0	0	3	100
Total	67	29	164	71	73	32	152	68

Table 9.5 Age group and quality: single-family residential units

		Value ratio index				Housing quality index		
Age building groups	Needs rehab	Percent	Adequate	Percent	Needs rehab	Percent	Adequate	Percent
17	0	0	4	100	0	0	4	100
18	0	0	7	100	1	14	6	86
19	0	0	5	100	1	20	4	80
20	1	4	21	95	2	9	20	91
21	15	25	46	75	19	32	41	68
22	10	53	9	47	10	53	9	47
23	9	56	7	44	3	20	12	80
24	5	38	8	61	3	23	10	77
25	0	0	2	100	0	0	2	100
Total	40	26	109	73	39	26	108	74

rehabilitation are quite similar for both indices (value ratio index = 34% and housing quality index = 32%). Four age groups presented less than five units.

Of the types of housing units present in the area, multifamily units were the fewest, only 61 units for the value ratio index and 50 for the housing quality index. This number represents only 17% of all units assessed by the indices.

As a summary, in some instances, the composite indicators detected the same proportion of units considered inadequate. This was the case

Table 9.6 Age group and quality: duplex and triplex units

		Value ratio index				Housing quality index		
Age building groups	Needs rehab	Percent	Adequate	Percent	Needs rehab	Percent	Adequate	Percent
15	1	33	2	67	0	0	1	100
17	0	0	1	100	0	0	1	100
18	0	0	4	100	1	25	3	75
19	0	0	5	100	0	0	5	100
20	3	27	8	73	4	36	7	64
21	10	21	38	79	19	40	29	60
22	12	46	14	54	7	29	17	71
23	17	47	19	53	14	39	22	61
24	8	47	9	53	2	12	15	88
25	1	100	0	0	0	0	1	100
Total	52	34	100	66	47	32	101	68

Table 9.7 Age group and quality: multifamily units

		Value ratio index				Housing quality index		
Age building groups	Needs rehab	Percent	Adequate	Percent	Needs rehab	Percent	Adequate	Percent
12	1	100	0	0	0	0	1	100
14	10	50	10	50	1	8	12	92
15	8	73	3	27	0	0	9	100
16	1	50	1	50	1	50	1	50
18	1	100	0	0	0	0	1	100
19	2	67	1	33	1	50	1	50
21	4	67	2	33	2	40	3	60
22	3	60	2	40	2	40	3	60
23	7	63	4	36	5	45	6	54
24	1	100	0	0	0	0	1	100
Total	38	62	23	38	12	24	38	76

for the total of units when considering both owner-occupied and rental together, rental units, single family, duplexes, and triplexes. Where the proportions were different was for owner-occupied only and multifamily. It is important to note the low number of units present in these two categories (owner-occupied represented 23% and multifamily 17%). This lower number could explain the discrepancies in the estimates for both indices.

9.4.1 Testing the Significance of Differences Across Age for Units

Although the proportion of housing units considered inadequate appears to be different across age groups, it was necessary to check if that difference was statistically significant. For that purpose, a one-way ANOVA analysis was performed.

The test could only be applied to the value ratio index due to the level of measurement (scale/ratio) of this index. The one-way ANOVA is a robust test when analyzing indicators with these characteristics. The housing quality index was assessed, but presenting the details of the results goes beyond the aim of this paper.[10] On the other hand, the results of testing the value ratio index allowed us to contribute to the literature using this indicator. Furthermore, as we will

[10]A Kruskal-Wallis test (a nonparametric test usually deemed fit for an ordinal level of measurement) was conducted to examine the difference between age groups by using the original eight scores of the housing quality index. Significant differences ($H = 25.03$, $p = 0.01$, $df = 12$) were found between the age groups.

see in the following analysis, it will be more important to test the relationship between both indices (and their correspondence with code violations) to see if they are estimated in a similar way independently of the age groups.

The following results are presented in a general and disaggregated manner to show the results based on type and occupancy of units.

The one-way ANOVA result for both owner-occupied and rental units shows a statistically significant difference (see Table 9.8) between group means (F (9,291) = 7.06, p = .000). A Tukey post hoc test revealed that there was a statistically significant difference among the age groups from 18 through 22.

Figure 9.4 shows the plot of mean values, which clearly depicts the decrease in the mean value for the value ratio index as the units get older. The plot also suggests where the differences between groups lie. The behavior of age groups 15 and 17 is explained by the low number of units in these two groups (3 and 5, respectively), which limits the test to find out differences.

The one-way ANOVA results for both separate owner-occupied and rental units show a statistically significant difference (see Tables 9.9 and 9.10) between group means (rental units, F (9,221) = 5.59, p = .000; and owner-occupied, F (7,62) = 8.95, p = .000). A Tukey post hoc test revealed that there is a statistically significant difference among the age groups from 18 through 22 in the case of rental units. There are no Tukey test results for owner-occupied because of the low number of cases in each group. The plot of means for rental units (Fig. 9.5) shows where the differences may lie.

Tables 9.11, 9.12, and 9.13 show the one-way ANOVA results regarding type of housing units. Single-family residential (SFR) and duplex and triplex (DT) units show a statistically significant difference between group means (SFR, F (8,140) = 7.09, p = .000; and owner-occupied, F (9,142) = 307, p = .002). In the case of multifam-

Table 9.8 One-way ANOVA – owner-occupied and rental units

	Sum of squares	df	Mean square	F	Sig.
Between groups	1.014	9	.113	8.480	.000
Within groups	3.867	291	.013		
Total	4.881	300			

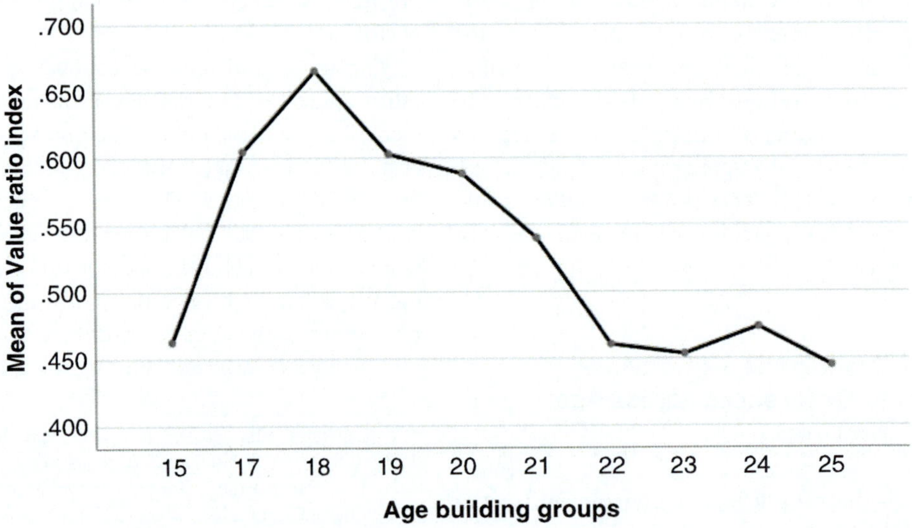

Fig. 9.4 Mean plot – owner-occupied and rental units

Table 9.9 One-way ANOVA – rental units

	Sum of squares	df	Mean square	F	Sig.
Between groups	.682	9	.076	5.592	.000
Within groups	2.994	221	.014		
Total	3.676	230			

Table 9.10 One-way ANOVA – owner-occupied units

	Sum of squares	df	Mean square	F	Sig.
Between groups	.606	7	.087	8.959	.000
Within groups	.599	62	.010		
Total	1.206	69			

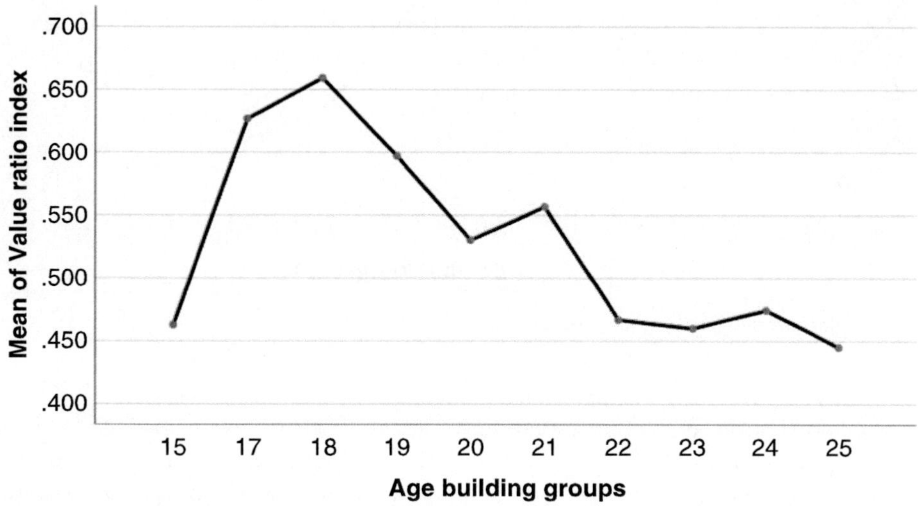

Fig. 9.5 Mean plot – rental units

Table 9.11 One-way ANOVA – single-family residential units

	Sum of squares	df	Mean square	F	Sig.
Between groups	.731	8	.091	7.093	.000
Within groups	1.804	140	.013		
Total	2.535	148			

Table 9.12 One-way ANOVA – duplexes and triplexes

	Sum of squares	df	Mean square	F	Sig.
Between groups	.369	9	.041	3.072	.002
Within groups	1.896	142	.013		
Total	2.265	151			

Table 9.13 One-way ANOVA – multifamily

Sum of squares		df	Mean square	F	Sig.
Between groups	.111	9	.012	.931	.507
Within groups	.677	51	.013		
Total	.788	60			

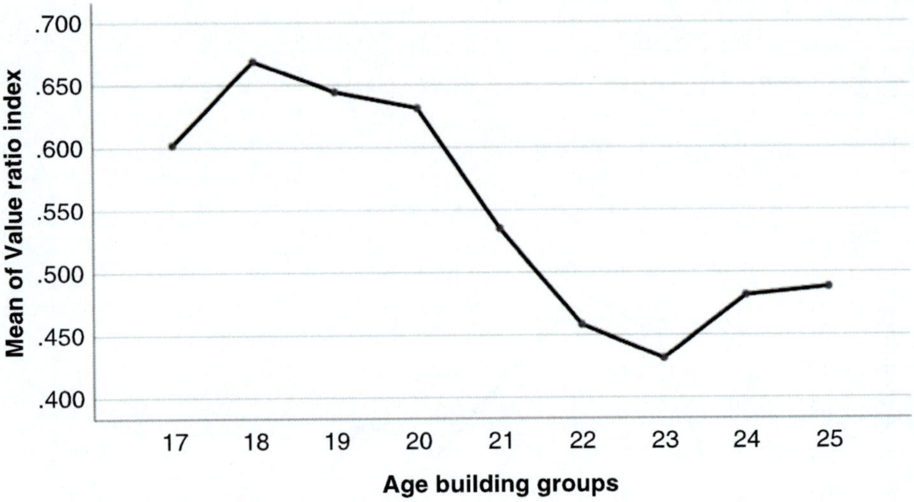

Fig. 9.6 Mean plot – Single-family residential units

ily, there was no significant difference between groups (F (9,51) = .93, p = .507).

A Tukey post hoc test revealed that there is a statistically significant difference among the age groups from 18 through 24 in the case of single-family residential units. There are no Tukey test results for duplexes and triplexes and multifamily because of the low number of cases in each group. The plot of means for single-family residential and duplex and triplex units (Fig. 9.6) shows where the differences may lie (Table 9.11).

9.4.2 Comparing the Two Composite Indicators

Comparing the two indices is essential in order to check for correspondence and to detect areas for improvement in their creation. So far, the previous analysis has shown that both indicators offer similar results in identifying the total number of inadequate housing units. In this section chi-square[11] statistics were obtained to assess the statistical significance of the correspondence between the value ratio and the housing quality indices. The assumption is that if both indicators are measuring inadequate housing accurately, then there will be a significant statistical correlation between them.

Like previous analysis, results are presented by occupancy and type of housing units. In this way, the results could be explained, considering where the similarities or differences could be.

The overall correspondence between the two composite indicators (considering owner-occupied and rental units together) was not statistically significant, $X2$ (1, N = 295) = .72, p > .05, even though the results in number of total units from both indices are quite similar (30% and 29%; see Table 9.2). This result could be due to the discrepancy in the results when looking at the results by the type of occupancy. The chi-square

[11] Also known as a chi-square test of independence.

tests for both separate owner-occupied and rental units show that the correlation of the indices is statistically significant in the case of owner-occupied units, $X2$ (1, $N = 70$) = 7.81, $p < .05$, and it was not significant for rental units, $X2$ (1, $N = 275$) = .31, $p > .05$. The reason for this could be the overall age of buildings in the study area. The age of buildings goes beyond the parameters examined in the literature. Older buildings start either holding their value due to renovations or loosing value due to deterioration; thus, the value ratio index does not detect these differences.

In terms of the correspondence between the indicators regarding the types of housing, only on single-family residential units was the relationship between the indices significant, $X2$ (1, $N = 147$) = 2.79, $p > .05$. The relationship with duplexes and triplexes and multifamily units were not significant, $X2$ (1, $N = 148$) = .19, $p > .05$, and $X2$ (1, $N = 50$) = .01, $p > .05$, respectively. Like the conclusion above, this could be due to the specific age conditions of the units and smaller numbers in the case of multifamily buildings.

Finally, though no less important, the overall relationship between the housing quality index and code violations was statistically significant, $X2$ (2, $N = 278$) = 7.24, $p < .05$. This result validates the creation of the composite indicator when relying on data from assessment of external conditions. As expected, the value ratio index's statistical correlations with code violations are not significant. This was expected, as the index is calculated based primarily on considerations of age.

In general, although the statistical correlation between the housing quality index and records of code violations is significant, the indicator could be adjusted to consider some elements of the housing stock to produce better composite indicators, for instance, considerations due to the nature of the way traditional single-family residential units have been transformed into places that offer rooms to students.

9.5 Conclusions

Two composite indicators are proposed with the goal of measuring inadequate housing at the neighborhood level. One index identifies inadequate units by considering physical exterior conditions and the other index by using the value ratio of units based on information contained in administrative records from tax assessments.

The results of the two composite indicators show an overall similar identification of housing units deemed to be inadequate. Both indices work well in tandem, allowing both a detailed analysis of housing conditions and a prompt analysis to identify how many units have deteriorated and at what time they begin to do so.

The results of the procedures aiming to test the statistical significance of the relationship between age and inadequacy show that there is a significant relation between both composite indicators and the age of housing units. This validates what previous authors have found about this relationship.

Although it seems not to be a statistically significant overall correspondence between the two indices, it was found that there is a statistically significant correlation when looking at the relationship between the composite indicators and the results for owner-occupied and single- family residential units. The physical characteristics of rental and multifamily units could make it difficult to assess their physical conditions in the field. This difficulty is common in the literature on measuring inadequate housing.

The statistically significant relationship between the housing quality index and records for code violations validates not only the index but also recent findings in the literature about the relevance of exterior conditions in measuring inadequate housing.

From a housing planning perspective, it is important to point out that the presence of deteriorated or inadequate housing in this neighborhood could well be explained by the large proportion of rental units and type of renters. As the literature shows, this is an important determinant of quality of housing. Students may be more interested in just having a place to dwell while they finish their studies and therefore do not worry too much about the quality of their housing units.

At this point, the indicator obtained is considered in the process of updating the neighborhood's master plan. It is imperative that this indicator can be used to track progress in the implementation of SDGs in the community.

Replication of the proposed methodology in this case study is paramount so that other communities can create data for evidence-based policymaking; thus, some generalizations and recommendations follow.

9.5.1 Generalizations and Recommendations

The ubiquity of the data used in this study in the United States allows us to conclude that other communities can easily replicate the proposed methodology and create their own indicators of inadequate housing.

The experience in this case study allows us to confirm that the data collection methodology used is quite suitable for adoption and implementation by resource-strapped neighborhood organizations interested in reporting on the SDGs.

Finally, two recommendations can be made: First, communities can use the methodology proposed to assess housing conditions and create an indicator of inadequate housing as part of an evidence-based policymaking process to implement goal 11 of the SDGs. Second, this methodology should be re-implemented every other year to capture change and to keep track of goal 11 of the SDGs, keeping in mind that the nature of the market and context (predominantly rental or owner-occupied) are important determinants of the ideal frequency.

Acknowledgments The author gratefully acknowledges the valuable contributions of students from the Geography and Planning Department at St. Cloud State University, City of St. Cloud officials, the Neighborhood University Community Council, and the Southside University Neighborhood Association.

References

Bianchi SM, Farley R, Spain D (1982) Racial inequalities in housing: an examination of recent trends. Demography 19(1):37–51. https://doi.org/10.2307/2061127

Eggers F, Moumen F (2013) American housing survey: housing adequacy and quality as measured by the AHS. U.S. Department of Housing and Urban Development, Washington, DC

Emrath P, Taylor H (2012) Housing value, costs, and measures of physical adequacy. City 14(1):99–125

Estevez L (2019) Composite indicators and data of housing assessment for St. Cloud, MN (Version V1): Harvard Dataverse. Retrieved from https://doi.org/10.7910/DVN/6HBLL7

Kain JF, Quigley JM (1970) Measuring the value of housing quality. J Am Stat Assoc 65(330):532–548. https://doi.org/10.1080/01621459.1970.10481102

Koebel CT (1986) Estimating substandard housing for planning purposes. [Article]. J Plan Educ Res 5(3):191–202. https://doi.org/10.1177/0739456x8600500306

Kristof FS (1968) Urban housing needs through the 1980s: an analysis and projection. Research Report No. 10, prepared for the National Commission on Urban Problems, Washington, D.C

Kutty NK (1999) Determinants of structural adequacy of dwellings. J Hous Res 10(1):27

Levine (1978) *Federal housing policy: current programs and recurring issues*. A background paper of the congress of the United States congressional budget office. U.S. Government Printing Office, Washington, DC

Newman SJ, Schnare AB (1988) Subsidizing shelter: the relationship between welfare and housing assistance. The Urban Institute Press, Washington, DC

Newman SJ, Struyk RJ (1983) Housing and poverty. Rev Econ Stat 65(2):243–253. https://doi.org/10.2307/1924490

Orr JA, Peach RW (1999) Housing outcomes: an assessment of long-term trends. Fed Res Bank New York Econom Policy Rev 5:51–61

Simonson J (1981) Measuring inadequate housing through the use of the annual housing survey. U.S. Department of Housing and Urban Development, Washington, DC

Sinha RC, Sarkar S, Mandal NR (2017) An overview of key indicators and evaluation tools for assessing housing quality: a literature review. [Review]. J Institut Eng (India): Series A 98(3):337–347. https://doi.org/10.1007/s40030-017-0225-z

Smith MT, Murray MS, O'Dell W (2003) Estimating the need for single family rehabilitation. [Article]. J Reg Anal Policy 33(2):37–52

Sumka HJ (1977) Measuring the quality of housing: an econometric analysis of tax appraisal records. Land Econ 53(3):298–309. https://doi.org/10.2307/3146122

Sutermeister O (1969) Inadequacies and inconsistencies in the definition of substandard housing. In: *Housing code standards: three critical studies*. Research Report No. 19, prepared for the Consideration of the National Commission on Urban Problems, Washington, DC, pp 78–105

U.S. Census Bureau (2017) Highlights of annual 2017 characteristics of new housing. Retrieved from https://www.census.gov/construction/chars/highlights.html

Watson NE, Steffen BL, Martin M, Vandenbroucke DA (2017) Worst case housing needs 2017 report to congress. U.S. Department of Housing and Urban Development, Office of Policy Development and Research, Washington, DC

Weicher JC (1979) Urban housing policy. In: Straszheim MA (ed) *In current issues in urban economics*. Johns Hopkins University Press, Baltimore

Weicher JC (1980) Housing: federal policies and programs. American Enterprise Institute for Public Policy Research, c1980, Washington, DC

Zwick PD, Schneider RH (1990) Searching for substandard housing in Alachua county: a case study. Comput Environ Urban Syst 14(4):273–282. https://doi.org/10.1016/0198-9715(90)90002-B

Luis Estevez, PhD, is a professor in the Department of Geography and Planning and the director of the School of Public Affairs Research Institute at St. Cloud State University. He is a graduate of the National Autonomous University of Mexico. He is a Fulbright scholar and holds a Master degree and PhD in urban and regional science from Texas A&M University. His research and experience focus in comprehensive planning, housing, land-use regulations, and assessment of indexes.

10 How the Youth Social Entrepreneurship (YSE) Model Supports Implementation of SDG #8 to bring Economic Growth and Decent Work in American Cities

Peg Thomas, Nancy Jacobs, Jennifer Valorose, and Laura Schauben

10.1 Introduction. The UN Sustainable Development Goals and the YSE Model

The United Nations Sustainable Development Goals (SDGs) are a framework of 17 goals designed to uncover and aggregate information in a manner that provides comparisons between disparate global areas and sectors. Indicators to track progress toward each goal are defined through 169 global monitoring indicators with suggestions for Complementary National Indicators. While used throughout Europe and Africa, these indicators have yet to be widely used in the United States and by nonprofits working to change public policies (Fig. 10.1).

The SDG #8 Workforce Development goals promote sustained, inclusive, and sustainable economic growth, full and productive employment, and decent work for all. Two of the subgoals, 8.5 and 8.6, and strategy 8.B.1 are focused on operationalizing national strategies for youth employment toward full and productive employment and decent work for all women and men, including young people and persons with disabilities, and equal pay for work of equal value (Fig. 10.2).

Amy Liu (2018), President and Director of Brookings Metropolitan Policy, examined how the 17 SDGs can help American cities tackle urgent local economic, political, and environmental challenges vital to the health and well-being of their residents. Liu concludes that the SDGs "reflect the imperative to create an economy of opportunity that promotes economic mobility for all." Elli Anzilotti (2018) notes that there are significant barriers for American cities including "poverty, racial inequity, climate inaction, and failure to provide healthy food to all residents" which make implementing common indicators challenging. While these issues are not unique to cities in the United States, Anzilotti states that there are certain aspects of American politics and culture that render them especially difficult to overcome. Anzilotti cites Jessica Espey, Senior Advisor for the Sustainable Development Solutions Network, who states, "Racial inequity is something we identified as a theme in every American city. When looking at differences between white and nonwhite groups in cities across America… there are acute differences in many outcomes–from poverty to health outcomes" (Anzilotti 2018).

The SDG Index and Dashboards Report for the Organization for Economic Co-operation and Development (OECD) countries shows that many

P. Thomas (✉) · N. Jacobs
Sundance Family Foundation, St. Paul, MN, USA

J. Valorose · L. Schauben
Wilder Research, St. Paul, MN, USA

D. B. Abraham, S. D. Iyer (eds.), *Promoting the Sustainable Development Goals in North American Cities*, Sustainable Development Goals Series, https://doi.org/10.1007/978-3-030-59173-1_10

Fig. 10.1 Venn diagram of the three domains of a YSE

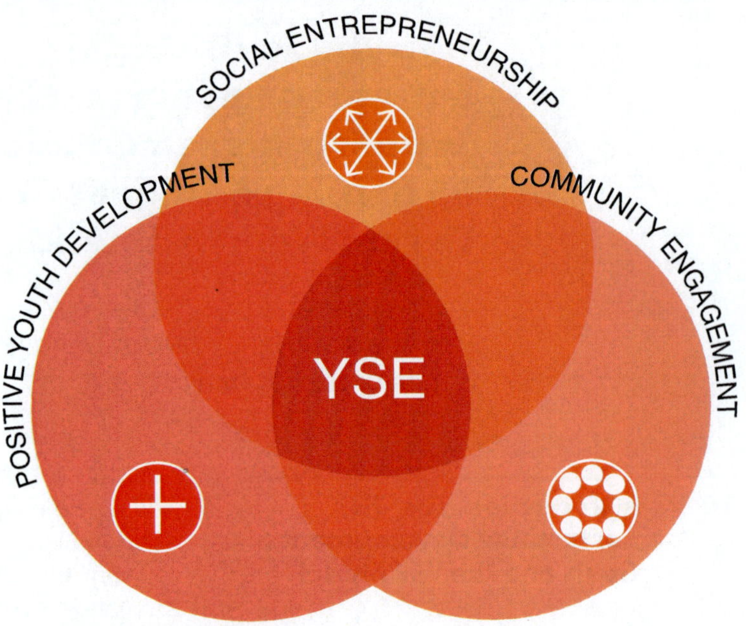

SDG 8 Decent work and economic growth

8.5 by 2030 achieve full and productive employment and decent work for all women and men, including for young people and persons with disabilities, and equal pay for work of equal value

Indicators include:

Youth employment rate, by formal and informal sectors and

Ratification and implementation of fundamental ILO labor standards and compliance in law and practice.

8.6 by 2020 substantially reduce the proportion of youth not in employment, education or training

Indicators include:

Secondary completion rates for girls and boys
Tertiary enrollment rates for women and men
Youth employment rate, by formal and informal sector

8.B By 2020, develop and operationalize a global strategy for youth employment and implement the Global Jobs Pact of the International Labor Organization

8.B.1 Existence of a developed and operationalized national strategy for youth employment, as a distinct strategy or as part of a national employment strategy

Fig. 10.2 SDG goals, indicators, and strategies 8.5, 8.6, and 8.b.1

are experiencing low economic growth and high unemployment (SDG 8), as well as major shortfalls in gender equality (SDG 5). This includes the United States, and the Metropolitan Statistical Areas that include American cities (SDSN 2017). Efforts to deploy youth-focused employment strategies which include trainings, supports, and skills development like the YSE multi-indicator/multi-sector model may help to contribute both to a fuller SDG dashboard and the data needed to inform local and regional policy.

Minnesota teens and young adults, in the Twin Cities metropolitan area, and in particular youth from communities of color, or youth living in low-income census tract areas, face the largest disparities in the United States. These disparities are dramatic: Minnesota is listed as having the fourth lowest overall poverty rate in the country at 10%. And yet, in 2017 and continued through today it has a high overall rate of opportunity youth[1] at 6.2% (Lewis 2019) with 27.1%, American Indian youth, and 12.1% of Black youth not in school or in work. Minnesota is also listed as having the worst financial and economic disparities in the country between white and Black families landing in 51st place in a national states survey. Three years later, Minnesota barely moved from 51st place to 50th in a states survey by *The Wall Street Journal*. These disparities and the gaps between Black and white Labor Market Participation Rate (LMPR) and the Employment to Population Ratio (EPOP), which have been tracked since 1972, reveal the need for a systemic solution (Ajilore 2020). A call for a multi-indicator/multi-structured model is needed to address the complexity of the issue.

In 2016, the Sundance Family Foundation partnered with Wilder Research to engage a cohort of youth-serving nonprofits that supported Youth Social Entrepreneurship (YSE) programs in a 36-month evaluation, capacity building, and research study. Sundance defines YSE programs as youth-designed, youth-led enterprises (i.e., businesses or community social initiatives) that include (1) social-emotional learning (SEL), (2) community and cultural engagement, and (3) the development of entrepreneurial thought, business, and work-readiness skills (Fig. 10.1). The outcomes include increased youth personal agency, enhanced social capital, and portable skills with a path to twenty-first-century careers. These youth-serving nonprofits typically have an embedded enterprise: youth repair bicycles, build boats, make cookies, sell recycled clothing, raise and sell vegetables, or a combination of these enterprises. While all nonprofits in this cohort share this common model, few had been producing quantitative data that would support this model as evidence-based.

The purpose of the study was to (1) grow the capacity of YSE nonprofit programs, (2) conduct evaluation, and (3) gather evidence of YSE model's impact with youth. As part of the common evaluation, Wilder Research implemented two surveys for all programs: an existing survey from the US Department of Labor for use by supervisors to capture youth's work readiness abilities and a newly developed survey focusing on youth personal agency (Wilder 2019).

While conducting the study, the researchers identified the indicators that matched the indicators of the United Nations Sustainable Development Goals (SDGs) and in particular align with the SDG #8—Workforce Development. The lessons learned over the 36 months of implementing common indicators across many different types of YSE programs will inform the implementation and replication of the SDG's workforce indicators, and strategy 8.b.1 to create a specific youth workforce development system by nonprofits serving youth in other American cities.

The benefit to the Twin Cities and other American cities to develop a cohesive set of measurable indicators, such as SDGs, is that overall improvements can be more successfully monitored and can be used to activate program or policy changes at a local or regional level.

[1]Youth who are neither in school nor the workplace.

10.2 The Youth Social Entrepreneurship (YSE) Model

The Youth Social Entrepreneurship (YSE) model integrates the practices of positive youth development with community engagement and social entrepreneurship to enable mutual transformation of economies, neighborhoods, and individuals. It was inspired by Dr. Melvin Delgado (2004) and further explored in a white paper prepared by Dr. Tina Kruse, entitled *Youth Social Entrepreneurship: Advancing the Field*, for the Sundance Family Foundation (Kruse 2015), and in a book published by Oxford Press (Kruse 2019).

Youth Social Entrepreneurship offers a unique opportunity for youth by bringing three key components of youth asset development strategies together, as described further below:

10.2.1 Component A. Social-Emotional Learning and Forging Youth-Adult Relationships

Focusing on social-emotional development provides youth a structured means to improve interpersonal skills. "The interconnectedness that results from social youth enterprises brings youth centrally into their own communities and does so in a manner that lends itself to creation of youth-adult partnerships and interactions that historically have not been present in this society" (Delgado 2004).

The YSE model supports the development of personal agency. This became one of the strongest motivators for the development of new data tools in the Wilder Research study by the YSE managers as described later.

10.2.2 Component B. Community and Cultural Engagement to Create Social Capital

Focusing on community and cultural engagement supports youth as they explore their own cultural environments, take progressive leadership roles, and participate in civic and cultural opportunities within their communities. As the youth become more culturally engaged, their social capital increases. Social capital can be helpful in finding and maintaining a job, a family, and a home (Woolcock and Narayan 2000). Helping youth cultivate their social capital and networks is admittedly a very long-term strategy but one that can help young people to create predictability and gather resources as described in *Bowling Alone: The Collapse and Revival of American Communities* by Robert Putnam (2000).

Because parental social capital has been shown to be an important predictor of adolescent social capital, it follows that youth whose families are marginalized may experience lower levels of social capital. To this end, YSE programs can enrich youths' social networks by supporting interactions with broader sets of individuals. Facilitated development of relationships could thereby grow an individual young person's social networks, eventually affecting their future opportunities (Weiss, 2012).

10.2.3 Component C. Work Readiness and Entrepreneurial Thought

Focusing on interdisciplinary elements of workforce readiness, the concept of entrepreneurship is taught through explorative value creation (Kruse, 2019). Youth in YSE enterprises are typically creating financial plans, dealing with customers, and communicating a vision (Fisher et al. 2008). These activities enforce the development of agile thinking, team interactions, and completing assignments both alone and together with others, as well as taking direction and providing peer feedback to others.

In traditional entrepreneurship education programs, workforce readiness might be taught through exercises, or role-playing in hypothetical situations. Using the YSE model, students participate in various aspects of the enterprise. For example, they may work as teams and often take independent roles after observing their peers. While this makes the enterprise inefficient in

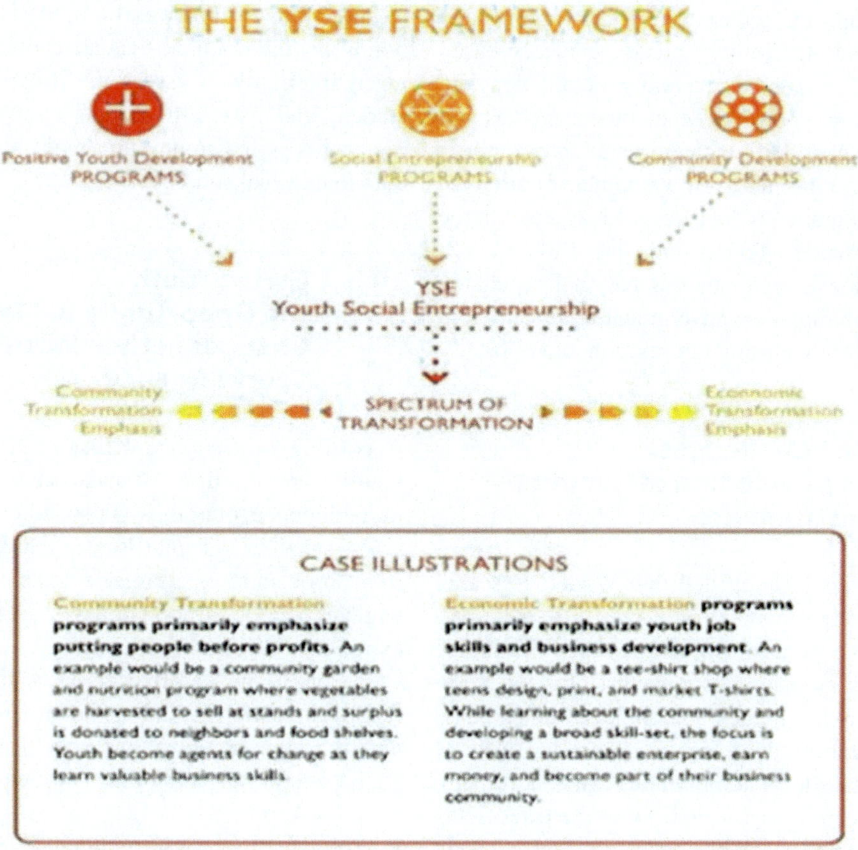

Fig. 10.3 The YSE framework to impact community and economic transformation

terms of work output, or a financial return on investment (ROI), it creates a learning lab where youth can naturally apply learned skills toward added value. The duration of these trainings varies, but a duration of more than 90 days is optimal (Fig. 10.3).

Building intergenerational relationships, as this model does, often involves partnerships, mentorships, or mutual experiences between a young person, volunteer, staff member, community member, and/or customer who is older by 5 years or more. These key relationships allow the youth to find resources and support from within their communities. Valuing their communities contributes to both community transformation and economic transformation. Community transformation programs emphasize people before profits, where youth might be paid as they create community gardens or save heritage seeds for the community. Economic transformation programs foster youth job skills and business development. These are the primary programs in the Twin Cities and are found in youth-led nonprofit enterprises.

"With such profound potential for positive impact, the YSE model can be viewed as an independent approach to youth development deserving of the empirical, practical, and financial support afforded other youth initiatives" (Weiss 2012).

10.3 The Sundance Cohort's Evidence-Based Study of the Youth Social Entrepreneurship Model

Inspired by the Social Initiative Fund (SIF) in 2015, the Sundance Family Foundation engaged with Wilder Research to design an evaluation that would enable nonprofits focused on youth training and development to jointly create evidence-based research. The Sundance Family Foundation had been funding workforce development models and strategies since 2003 including a Wunderkammer (reverse job fair) and pathways to help youth from training programs. Wilder Research as a part of the Amherst Wilder Foundation has had a well-deserved local, regional, and national reputation helping communities, government units, and nonprofits thrive by turning information into impact.

10.4 The Challenges of Developing of Common Instruments

The Sundance study led by Wilder Research included nonprofits that were very different enterprises one to another—from urban agriculture, bike shops, bakery, and consignment stores to internships and micro-grants. This spoke to the variation of organizations working to impact the lives of youth in their local communities. All cohort members had enterprises where youth were receiving stipends or wages while working, and all were employing youth living in low-income areas or in communities of color, often in the neighborhoods where these nonprofits are located.

Initially in 2015, each of the cohort organizations saw themselves primarily engaged in youth development and not workforce career readiness or training. The cultural, community, and social-emotional learning activities varied greatly from one organization to another. Finding commonality took many meetings and several additional months. It was also important to all cohort program staff, not only to protect the youth from too much invasive questioning but to also measure indicators of social-emotional development and personal agency. How to do both at the same time created many engaging conversations that ultimately built trust between organizations, the foundation and the researchers. All nonprofits were using the Youth Program Quality Assessment Tools (YPQA) to ensure that their environments were youth sensitive, embracing, and free from gender and racial barriers. All nonprofits were reporting to countless government units and providing grant evaluations to funders. Yet, these nonprofits did not have the data captured or organized internally to assess the progress of each student and to retain records over time. Most nonprofits relied upon spreadsheets or reported data to external sources.

10.5 Giving Youth the Opportunity to Make Changes in their World Sparks Agency

According to Dr. Tina Kruse (2019), "Giving youth opportunities to make change in their immediate surroundings via social entrepreneurship—whether it's small-scale change in a classroom procedure or a larger scale enterprise that engages a neighborhood or beyond—results in an experience of genuine agency."

Through an iterative process of gathering feedback, cohort members identified collective and, for a few organizations, specific measures of social-emotional development which included:

- Youth are more self-confident.
- Youth set future goals/have future aspirations/future planning.
- Youth have increased interpersonal, teamwork, and leadership skills.
- Youth have increased sense of cultural identity.

Indicators measuring community and social connectedness included:

- Youth have positive relationships with caring adults in community.
- Youth feel more connected to each other.
- Youth feel more connected to the community.

Indicators to track improved lifestyles and healthy behaviors included:

- Youth have agency.
- Youth have improved financial self-sufficiency.
- Previous youth offenders do not reoffend; unoffenders do not start offending.*

- Youth increase bike riding as transportation.*
- Youth increase healthy eating.*

Note: Indicators marked with an asterisk (*) were not collected across all nonprofits.

After a 4-month process of surveys, consensus building, and refining logic models for each organization, Wilder Research created (1) a Youth Retrospective Survey to capture recommended indicators of personal agency. This survey gives youth a way to voice their own personal development consistent with increasing their social-emotional learning. Wilder Research also identified (2) an existing Work Readiness Tool, developed by the US Department of Labor in 1998 and revised in 2014. These two surveys and demographic data captured about each participant formed the basic data for this study. The Workforce Readiness survey is administered by a work-site supervisor and assesses youth on the following (definitions for each measure are included in the tool): (1) attendance, (2) punctuality, (3) workplace appearance, (4) taking initiative, (5) quality of work, (6) communication skills, (7) response to supervision, (8) teamwork, (9) problem-solving/critical-thinking, and (10) workplace culture policy and safety. Three additional items were added at the request of the group: (11) involvement in something larger than themselves, (12) financial self-sufficiency, and (13) job application/employability skills.

10.6 Preparing to Implement a Common Set of Indicators

Implementing these common indicators was a challenge for a variety of reasons. Youth-serving nonprofits most often direct their financial and personnel resources to implementing programs and not capturing data. Adding these indicators, even with a variety of data capturing tools, required training, the use of new or revised forms, and new dashboards or analysis. Capturing a common set of indicators was enhanced with the following:

Selecting an internal team of three key staff from each organization and asking all three to attend several training meetings were important. Two to three large group meetings each year both built teams in each organization and offered networking between organizations. The three staff from each nonprofit consisted of (1) an executive director or a data manager, (2) the grant writer, and (3) the direct program manager. As time passed, sending three key staff to trainings was not seen as cost-effective by the nonprofit leadership even when many organizations experienced staff turnovers exceeding 150% in the positions. Nonprofits started sending smaller teams. However, without these collective meetings, adherence to data collection and interagency networking may have waned. The fatigue which is always associated with learning and implementing new systems was assuaged to some extent by Wilder Research. To support implementation of common evaluation instruments, Wilder Research scientists were assigned to specific organizations and met with each several times between January and June 2016 to develop logic models and evaluation plans. Each cohort member arranged to meet in-person with their assigned research scientist to review the Work Readiness and Youth Retrospective tools, review the grading scales, and understand the measurement definitions prior to implementing the tools. The researchers stressed the importance of implementing the tools consistently.

Staff/supervisors were instructed to complete the Work Readiness Tool for each youth employee/participant approximately 2 weeks after their programming started and then approximately every 90 days thereafter. A youth in the program for a year would then be assessed 4 times over the course of that year.

Training instructions were also developed for cohort members moving to databases for the first time. Intensive in-house one-to-one trainings were required in many areas: moving from spreadsheets to a database to capture longitudinal data, training program staff to implement the surveys, and training a central data collector. Reports included blind templates for evaluators and templates with full data for staff. Some cohort sites opted to use Survey Monkey to capture their Youth Retrospective Survey results. Instructions for using the forms with youth were also carefully explained to each supervisor. These and other areas are typically not developed in organizations that are not affiliated with universities or research institutions. Developing the infrastructure to collect these indicators was more expensive than anticipated and required more time, training, and support staff.

10.7 Gathering Data Using Common Indicators from Un-common Youth-Serving Nonprofits

The commonality for all cohort members was that they had youth-driven enterprises and used the YSE model. These programs varied widely in terms of size and enterprise activity. Of the initial 14 programs (12 organizations) that participated in the YSE research study, 11 are included in this analysis. Programs vary: 6 have less than 25 staff members, while 2 programs have more than 100 staff members; 9 are located in Minneapolis with 2 in St. Paul, and 7 programs employ youth for more than 5 months. Due to variations in their programming and their initial data capacity, the number of youth assessed by each program as part of the evaluation varied substantially (4–196 youth) (Fig. 10.4).

From September 2016 through August 2018, programs implemented the group instruments in accordance with the protocols. As a result, across all programs, 46% of youth (n = 412) were assessed using the Work Readiness assessment, and 58% of youth (n = 519) completed the Youth Retrospective Survey.

10.8 Replication of the Study for Reliability

From September 2017 through August 2018, 11 YSE programs replicated the study with 889 youth. Data was limited to youth age 13–24. Programs range in size, serving between 12 and 286 youth during that 2-year period, with some offering only summer programming.

About half of youth were identified as male and half as female, with one program serving only women. Most youth (81%) were 14–17; all programs served this age range. The average age was 16. Twenty percent were young adults age 18–24, with most programs (8 of 11) serving this older group. Younger teens and Black youth are more represented in the Work Readiness results. This is in part due to the fact that one particular program that served older youth had difficultly completing their post-assessments.

All programs served racially diverse youth. About half of the youth served were identified as Black, African American, or African (54%) and were represented in all 11 programs. Youth next most commonly identified were White or Caucasian (18%), Asian or Asian-American (14%), American Indian or Alaska Native (5%), and Hispanic or Latinx (3%). The other youth represented a variety of racial groups, with the programs themselves serving different ethnic, racial, or cultural groups of low-income youth.

10.9 Outcomes

At the end of data collection in September 2018, 864 youth had participated in their respective YSE program up to 746 days (2 years), with an average of 114 days (3.7 months). During the 24-month period, staff at 10 of the 11 YSE programs assessed youth at least twice on 12 Work Readiness skills using the tool modified from the existing survey (US Department of Labor, 1998). A total of 412 youth were included in the analysis ranging from 8 to 152 youth per program.

Researchers limited the analysis of Work Readiness assessments to initial assessments that were completed between 1 and 30 days upon program entry so that scores would represent a true baseline. Youth with an initial baseline assessment of over a month after starting the program were excluded from this analysis. As a result, the youth (n=412) with a valid pre-assessment had

Fig. 10.4 Overview of the programs in the Sundance and Wilder Research YSE cohort

Size of programs (number of youth served annually)	Number
Small (less than 25)	6
Medium (25 – 99)	3
Large (100 or more)	2
Location	**Blank**
Minneapolis/Saint Paul	9
Suburbs	2
Type of enterprise	**Blank**
Retail (bakery, graphic design, clothing)	5
Repair shops (bikes, small engine)	3
Outdoor-related (gardening, farmers market, landscaping)	3
Length of programming	Blank
Short (<4 months)	4
Longer (5+ months)	7
Size of cohort engaged in data collection	**Blank**
Small (less than 40 youth evaluated)	7
Large (40 or more youth evaluated)	4

about half the number of days of programming of overall youth served.

Overall, there was statistically significant improvement in 11 of 12 work readiness skills measured (all except "attendance"). On average, youth (n = 308–407) were proficient on 6.6 items at pre-assessment, increasing to 9.6 items at post-assessment after 30 or more days of programming.

Youth (n = 407–412) showed the most improvement in four areas: (1) problem-solving/critical-thinking, with 32% becoming proficient and 45% demonstrating some improvement; (2) quality of work, with 36% becoming proficient and 52% demonstrating some improvement; and (3) communication skills with 31% becoming proficient and 45% demonstrating some improvement. While staff observed fewer youth improving in the

N=407-412	Percentage rated proficient or exemplary		Percentage with any improvement[b]	Percentage becoming proficient[c]
General professional skills	Pre	Post[a]		
Quality of work	43%	76%*	52%	36%
Workplace culture, policy, and safety	62%	87%*	43%	27%
Punctuality	64%	81%*	37%	22%
Workplace appearance	86%	95%*	30%	11%
Attendance	73%	79%	32%	16%
Interpersonal skills	**Blank**	**Blank**	**Blank**	**Blank**
Problem-solving/Critical-thinking	38%	68%*	45%	32%
Taking initiative	41%	69%*	46%	31%
Communication skills	48%	75%*	45%	31%
Teamwork	57%	83%*	47%	29%

Fig. 10.5 Staff ratings of work readiness skills, pre/post, and change of youth in programs in the Sundance Family Foundation and Wilder Research YSE cohort

Response to supervision	64%	86%*	46%	24%
Other (N=311)	**Blank**	**Blank**	**Blank**	**Blank**
Job application skills	46%	79%*	48%	34%
Financial self-sufficiency	46%	75%*	41%	30%

Note. Youth were rated on a 4-point scale: Exemplary, Proficient, Needs Development, or Improvement Plan Needed. Totals may not equal 100% due to rounding. To be included, pre-assessment had to be completed within 30 days of program entry, and the post-assessment had to be completed at least 30 days after the pre-assessment.

*[a] Significance tests were conducted using McNemars test; an * indicates the post was statistically higher than the pre.*

[b] Percentage becoming proficient are those youth that went from Needs Development or Improvement Plan Needed to Proficient or Exemplary.

[c] Percentage with any improvement are those youth who progressed to a higher category on the scale.

Fig. 10.5 (continued)

area of workplace appearance, a large proportion of youth (86%) were proficient or exemplary in this area initially with post scores of 95%. Attendance had no significant change (Fig. 10.5).

10.9.1 The Length of Involvement in a YSE Program Improved the Outcomes

Preliminary results indicate that the longer youth are employed, the more their work readiness skills may improve. On average, those engaged in a YSE program for less than 75 days (n = 180) were initially proficient on 7.0 items increasing to proficiency on 9.6 items at post-assessment. In comparison, youth (n = 126) engaged for more than 75 days were initially proficient on 6.1 items increasing to 9.7 items on the post-assessment.

Of note, 54% of youth with longer tenures improved on the indicators about their ***response to supervision*** and ***teamwork*** compared to 42–43% of youth with less duration. However, 35–37% of youth with shorter tenures were proficient on ***job application*** and ***financial self-sufficiency*** skills compared to 34–36% of those with more duration. Four programs engage youth on a shorter-term basis in early job and financial training, so these differences may be a result of programming differences between shorter-term and longer-term programs.

10.9.2 Youth Demographics Reveal Different Patterns of Acquiring Skills by Age and Gender

Researchers analyzed data based on youth demographics to determine whether certain groups benefit from YSE programs more than others. No clear patterns emerged; however, a few nuances are worth noting.

In comparing work readiness scores by gender, both young men and young women became proficient in a similar number of items. However, young women were initially generally more proficient. Though both young women and young men experienced similar levels of improvement, areas with initial statistically significant differences included:

- Young women (n = 161) were ***more likely*** to initially be rated proficient or exemplary on punctuality, workplace appearance, communication skills, and problem-solving/critical-thinking at the beginning of the program.
- Young men (n = 123) were ***less likely*** to initially be rated proficient or exemplary on indicators of punctuality; quality of work; workplace culture, policy, and safety; and problem-solving/critical-thinking at the beginning of the program, giving young men more room to improve.

10.9.3 Acquisition of Skills by Age of Youth

Youth age 16–17 experienced the most growth when compared to younger youth (age 14–15) and young adults (18–24). In part, youth age 16–17 were rated proficient on a fewer number of items initially, on average, meaning they had the most room to grow.

- Younger youth (n = 127 age 14–15) were ***more likely*** to initially be rated proficient or exemplary on indicators of punctuality; workplace appearance; quality of work; workplace culture, policy, and safety; and problem-solving/critical-thinking.
- Older youth (n = 105 age 16–17) were ***more likely to improve*** in attendance; punctuality; workplace appearance; quality of work; workplace culture, policy, and safety; and problem-solving/critical-thinking.
- Young adults (n = 34 age 18–24) were ***less likely to improve***; however they were ***more likely*** to demonstrate initial proficiency. Despite being proficient on 10 of 12 items after participating, in general, they were less likely to be proficient on indicators of workplace appearance, response to supervision, and teamwork.

African-American/Black and white youth in general demonstrated improvements in work readiness skills over the course of their involvement in their respective programs. Below are a few notable variations:

- Black and Asian youth were ***more likely to improve*** in workplace culture, policy and safety, and taking initiative.
- Asian youth were ***more likely*** to initially be rated proficient or exemplary on indicators of workplace appearance, quality of work, communication skills, and problem-solving/critical-thinking.
- White youth were ***more likely to decline*** on indicators of punctuality, workplace appearance, taking initiative, and response to supervision.

10.9.4 Impacts Reported on Social-Emotional Development

Across the 11 programs participating in the 23-month study, 519 youth completed the Youth Retrospective Survey. Youth rated their level of agreement or disagreement with each survey item twice: once based on how they felt at the initial time taking the survey, and once at the end of the program reflecting back to how they felt prior to joining their respective program.

Youth showed statistically significant growth across all four required survey items related to self-confidence, community connectedness, and goal orientation. A larger proportion of youth reported increases in feeling part of a community, with about half progressing to a higher category on the scale ("more agreement") and a quarter advancing from disagreement to agreement (Fig. 10.6).

Among the optional items (administered in 8–10 of the programs), a statistically significant change in youth's agreement was in all but one item. Of particular note, 40–50% of youth perceived growth on the following items, with about a quarter going from strongly disagree/disagree to strongly agree/agree:

- I have the skills and experiences needed to be a mentor for other youth.
- I feel comfortable speaking in front of a group of people.
- I know what I can do to help make the community a better place.

While less than 10% of youth indicated growth on the following four items, 90% or more of the youth initially agreed with these statements. This may indicate that the youth already perceived themselves as strong in those areas when they began their involvement with the YSE program:

- I think it is important to listen to and value the opinions of others.
- I believe young people can make a difference in the community.
- There are people in my life I can depend on when I need help.
- I am willing to stand up for what is right.

10.10 Key Learning Which May Influence How Common Indicators Are Placed upon Youth-Serving Nonprofits in American Cities

Changing the system of employment with youth while also imposing common indicators on organizations that may or may not have an existing data or analysis system can be daunting. This study tried to resolve some of these disconnects. Stumbling blocks included (1) significant turnover, often one, two, or three staff members every 18 months on the internal team; (2) the inexperience of program staff working with indicators to be used for public analysis and reporting; and (3) the lack of infrastructure to convert conversation into real data collection.

10.10.1 Key Overall Findings

YSE programs serve diverse youth. The youth served by the YSE programs were racially and ethnically diverse with an equal mix of young men and young women ranging in age from 13 to 24 (average age 16). Over half were identified as Black, African American, or African (54%).

Youth gained proficiency in work readiness skills. Overall, after 30 or more days of programming, staff observed growth in 11 of 12 work readiness skills measured (all except "attendance").

Youth perceived growth in their own social-emotional skills and personal agency. On the Youth Retrospective Survey, up to a third of the youth perceived improvement in their social-emotional skills. Youth rating their own skills at the beginning and end of programming reflected statistically significant growth in feeling part of their community, feeling confident, having an adult to talk to about problems, and having future goals for themselves.

	Percentage agree or strongly agree		Percentage with more agreement[b]	Percentage moving from disagreement to agreement[c]
Required items (11 programs)	Retro-Pre	Post[a]		
I feel like I am part of a community. (N=513)	73%	96%*	25%	48%
I am confident. (N=515)	72%	96%*	24%	43%
When I have a problem, there is an adult that I can talk to. (N=511)	78%	95%*	18%	38%
I have identified future goals for myself. (N=511)	75%	92%*	18%	35%
Optional items (8-10 programs)	**Blank**	**Blank**	Blank	Blank
I have the skills	64%	92%*	28%	51%
and experiences needed to be a mentor for other youth. (N=426)				
I feel comfortable speaking in front of a group of people. (N=426)	57%	81%*	24%	43%

Fig. 10.6 Youth self-ratings, retro-pre/post, and change, in 11 programs in the Sundance Family Foundation and Wilder Research YSE cohort

I know what I can do to help make the community a better place. (N=426)	70%	92%*	23%	45%
I can handle stressful situations. (N=426)	74%	93%*	20%	41%
I feel supported in pursuing my personal goals. (N=463)	84%	97%*	14%	39%
I know how to get along with other young people. (N=462)	86%	97%*	11%	38%
I believe young people can make a difference in the community.	89%	96%*	8%	29%
(N=426)				
There are people in my life I can depend on when I need help. (N=400)	90%	96%*	7%	26%
I am willing to stand up for what is right. (N=342)	92%	97%	6%	25%

Fig. 10.6 (continued)

I think it is important to listen to and value the opinions of others. (N=466)	95%	99%*	4%	30%

Note. Youth rated themselves on a 4-point scale from strongly disagree to strongly agree. Totals may not equal 100% due to rounding.

[a] *Significance tests were conducted using McNemars test; an * indicates the post was statistically higher than the retrospective-pre at* $P < 0.001$ *level.*

[b] *Percent moving from disagreement to agreement are those youth that went from strongly disagree or disagree to strongly agree or agree.*

[c] *Percent with more agreement are those youth who progressed to a higher category on the scale.*

Fig. 10.6 (continued)

Cultivating the workforce readiness and career paths requires the development of an articulated skills pathway, where a credential or nationally recognized skill achievement will simultaneously provide credits for a degree. The degree credits will contribute to a degree program and ultimately to a 2-year and then a 4-year diploma. The call for an accelerated effort to attain 70% post-secondary credential or degree attainment with adults age 25–44 makes this an excellent time to create a distinct strategy for a youth workforce development system. In 2015 the Minnesota Legislature enacted a state postsecondary educational attainment goal that 70% of Minnesota adults (age 25 to 44) will have attained a postsecondary certificate or degree by 2025 (Minn. Laws 2015 Chapter 69 Article 3 Sec. 6). Achieving this goal has been uneven and markedly slow (Hermida and Fergus 2019). Accelerating progress towards the goal makes this an excellent time to create a distinct strategy for a youth workforce development system as identified in SDG 8.b.1.

10.10.2 The Need for the Implementation of a Formal Youth Training Model to Inspire Collaborations to Collect Common Indicators Requires Coordination and Large Measurable Goals Such as the UN SDGs

SDG #8 measures both formal and informal workforce development. Informal models historically make up the informal economic sectors, which exist in many areas of the world and which are also found in American cities. Youth and other marginalized people have long engaged in entrepreneurship as a means of survival. As a result, in many areas "young people (and women of all ages) make up a significant portion of the informal sector due to the systemic disenfranchisement that limits their participation in formal employment" (Kruse 2019). Nonprofits and government units in American cities would do well to use formal models to create and capture common indicators.

The use of a formal model, like this YSE model, both cultivates and monitors youth developmental skill areas, which lead to career paths. The formalized structure also lends itself to application with various organizations so that common indicators can be used. When adapted, this YSE model may be a model that holds promise with youth work training programs throughout American cities and rural areas and in other OECD and non-OECD countries. Further study and replication is required.

10.11 Recommendations: Ten Lessons Learned in Creating a Formal Youth Workforce Development System Using the SDG Framework

1. Creating collaborations that include youth-serving providers, program designers and practitioners increases networking, professional development, trust, and cooperation. The resulting acceptance of common indicators coming from these nurtured relationships is of critical importance to the successfully implement of SDG strategies and indicators. It is very important that these collaborations be sustained to encourage the collecting and reporting of data for years after the initial training and implementation.
2. Buy-in to the acceptance of a set of common tools is a long process that must be curated. A concern in the field of youth work is that "many frontline staff begin with little training, and develop their professional skills in isolation… with limited opportunity to reflect, read research, or learn from peers or expert practitioners" (Larson et al. 2015). Aware of the fact that these grassroots nonprofits were growing in isolation from each other, in 2013 we developed a collaborative called the Youth Social Entrepreneurship Collaborative. This was a way to convene yearly summits, bring information about advancing the youth social entrepreneurship model, and allow organizational leaders to learn from one another. Workshops, speakers, social media discussions, and morning egg-bake breakfasts helped youth leaders gather to learn strategic topics and procedures.
3. Developing the infrastructure to collect these indicators was more expensive, time-consuming, and problematic than anticipated. Collecting data is often not viewed as effective and efficient programming by cash-strapped nonprofits serving youth in low-income areas or youth from communities of color. The ability of many nonprofits working with youth to collect yet another set of indicators is strained by a number of factors: financial, personnel, and back office capacity to capture and analyze data. The back offices for most community-based nonprofits may not include databases, or consistent personnel. A summer intern is often the designated accidental techie so that change in personnel is common and disruptive. These informal systems must be addressed and receive the capital needed before a collective system can be developed. While all programs funded with philanthropic or government dollars require evaluation, the requirements are not consistent. As a result, nonprofits managers, like managers everywhere, are understandably wary. The combination of working toward collective large-scale goals, such as creating a specific youth workforce development strategy, brings philanthropy, governments, and research to nonprofits to build capacity and capture data that will enable multiple partners to develop policy solutions. This work also enables nonprofits to document their successes, reach for larger funding, and ultimately track alumni and support them with innovations.
4. Putting youth first: nonprofit leaders are protective about burdening the youth they serve with yet more indicators. Many youth face daunting and invasive questionnaires in every part of their lives. Gaining trust from these youth requires that the youth have agency about their personal information.

All nonprofit leaders in this study were absolute in their decision to put youth first and to protect them from excessive analysis, data gathering, and surveys. It is important to listen both to the youth and to the nonprofit leaders as together they approve of the indicators and demographics that will both help the youth and the nonprofits.

5. Targeting improvements for opportunity youth and measuring progress using SDG 8.6 indicators may help communities more accurately find opportunity youth which enables the community, nonprofits, and local and regional governmental units to then provide solutions and support.
6. Creating a path of articulated skills development with appropriate supports at every level can guide youth from entry-level skills training programs to the acquisition of credentials and degrees at an earlier age, giving them more economic mobility.
7. Philanthropy must find a way to support the SDG common indicators: as the most charitable state in the nation (Burd-Sharps and Lewis 2013), Minnesota's philanthropy struggles to find common measures that can help guide its government policies and problem-solving solutions. Many efforts have and are being launched including this particular effort to create and implement protocols to capture common evidence-based data using a multi-indicator/multi-sector model. With expansion, it can galvanize the existence of a developed and operationalized strategy for youth employment as part of Minnesota's overall workforce development strategy.
8. Nationally, lessons from this study and the development of a YSE model may assist nonprofits serving low-income, opportunity youth and youth from communities of color in other American cities, by reinforcing the importance of youth-centered design.
9. While evidence-based challenges can and did arise during implementation of this YSE model, the lessons learned over the 36 months will inform the implementation and replication of the SDG's workforce indicators, and strategy 8.B.1 to create a specific youth workforce development system by nonprofits serving youth in other American cities.
10. Tying the implementation of SDGs to public policy will help secure its implementation and ultimate success. The benefit to the Twin Cities and other American cities to develop a cohesive set of measurable indicators is that overall improvements can be more successfully monitored and solutions can be coordinated at both a micro and macro level.

References

Ajilore O (2020) The persistent black-white unemployment gap is built into the labor market, The Center for American Progress Sept 28, 2020

Anzilotti E (2018) American cities are failing on sustainable development, Fast Company https://www.fastcompany.com/40582463/american-cities-are-failing-on-sustainable-development

Bertelsmann Stiftung and the Sustainable Development Solutions Network (SDSN) Sustainable Development Goals (SDG) Index and Dashboards 2017 edition

Burd-Sharps S, Lewis K (2013) One in seven: ranking youth disconnection in the 25 largest metro areas. Measure of America of the Social Science Research Council SSRC, New York

Compass Points (2017). https://www.mncompass.org/_pdfs/compass-points-2017.pdf

Delgado M (2004) Social youth entrepreneurship: the potential for youth and community transformation. Praeger, Westport

Fisher S, Graham M, Compeau M (2008) Starting from scratch: understanding the learning outcomes of undergraduate entrepreneurship education. In: Harrison RT, Leitch C (eds) Entrepreneurial learning: conceptual frameworks and applications. Routledge, New York, pp 321–356

Hermida A, Fergus M (2019) Educating for the Future. 2019 Update. Minnesota Office of Higher Education. - ERIC

Kruse T (2015) Youth social entrepreneurship: advancing the field. Sundance Family Foundation, Saint Paul. Retrieved from www.sundancefamilyfoundation.org/resources

Kruse T (2019) Making change: youth social entrepreneurship as an approach to positive youth and community development. NY Oxford Press, New York, p 109

Larson RW, Walker KC, Risk N, Diaz LB (2015) Understanding youth development from the practitioner's point of view; a call for research on effective practice. Appl Dev Sci 19(2):74–86

Lewis K (2019) Making the connection: transportation and youth disconnection. Measure of America, Social Science Research Council, New York

Liu A (2018) US cities leading on the sustainable development goals, NY brookings institution https://www.brookings.edu/blog/up-front/2018/12/19/us-cities-leading-on-the-sustainable-development-goals/

Putnam R (2000) Bowline alone: the collapse and revival of American communities NY. Simon & Schuster, New York

Watson DH, Lantos S, Villatoro H (2020) Rackers behind the numbers: what the latest opportunity index tells us about our communities August 27, 2020 The Form for youth Investment. https://forumfyi.org/blog/behind-the-numbers/

Weiss HE (2012) The intergenerational transmission of social capital: a developmental approach to adolescent social capital formation. Sociol Inq 82:212–235

Wilder Research (2019) Evaluation of twin cities youth social entrepreneurship programs: *Results from a point-in-time research study of 11 YSE programs funded by the sundance family foundation*

Woolcock M, Narayan D (2000) Social capital: implications for development theory, research, and policy. World Bank Res Obs 15(2):225–249

Peg Thomas Executive Director, manages the Sundance Family Foundation's operations, initiatives, and provides both grants management and technical support to grantees. Peg holds an MPA from Harvard Kennedy School, an MS in Therapeutic Recreation and Administration, an MPP in Business from the Carlson School of Business, and a BS in Music Therapy from University of Minnesota (with coursework from UConn, Napa Jr. College and University of California San Diego).

Nancy Jacobs holds an M.Ed. in Parent and Family Education from the University of Minnesota and a B.S. from Boston University. Nancy focused her early career in teaching early childhood development/parent education and teen parenting. Outside of directing the work of the foundation, Nancy is a business entrepreneur and serves as President of Metta Financial, LLC and a partner at Abby Investments, with a keen interest in expanding social enterprises, which include affordable housing.

Jennifer Valorose has over 15 years of research and evaluation experience primarily in the areas of early childhood education, youth development, and public health. She excels at partnering with nonprofit and government organizations to design and carry out tailored research projects to meet any organization's needs within small and large budgets. She is also skilled at building evaluation capacity through interactive trainings and individualized technical assistance, translating research findings into programmatic and policy recommendations, and presenting research findings in a variety of formats.

Her recent work includes several statewide child care studies for the Department of Human Services, evaluations of several out-of-school time programs, development of Minnesota's School Readiness Report Card, and being an evaluation consultant to several Department of Human Services – Behavioral Health Division grant initiatives. Her research interests include early childhood care and education, environmental health, home visiting, economic development, leadership development, and youth programming. Prior to joining Wilder Research in 2008, she worked for Rainbow Research conducting research projects on the following topics: housing and redevelopment, physical activity, tobacco cessation, and adult and youth education.

Laura Schauben has over 20 years of experience helping organizations achieve their goals through research and evaluation. Laura's primary areas of work include diversity, inclusion, and equity; mental and behavioral health; employment and community development; and trauma and violence. Current examples of her work include leading an assessment of the health, social, and human service needs of underserved communities in Minnesota, an evaluation of a statewide program promoting racial equity in employment, and an evaluation of entrepreneur programs that work to build financial equity in low-income neighborhoods across the country. Prior to coming to Wilder, Laura led an initiative addressing racial disparities in the criminal justice system and coordinated a counseling line for the LGBTQ+ community. Laura completed her M.A. in Psychology at the University of Minnesota and her B.A. in Sociology and Psychology at Brandeis University.

11 Local SDG Indicators for US Cities and Communities

David B. Abraham

11.1 Local SDG Goals and Indicators for the USA

Sustainable Development Goal indicators should be disaggregated, where relevant, by income, sex, age, race, ethnicity, migratory status, disability and geographic location, or other characteristics, in accordance with the Fundamental Principles of Official Statistics (General Assembly resolution 68/261).

End poverty in all its forms everywhere

Goals and targets (from the 2030 Agenda)	Adapted US local indicators – city or smaller geographies
1.1 By 2030, eradicate extreme poverty for all people everywhere, currently measured as families living on less than 30% of median family income for the metro area.	1.1.1 Proportion of population below the US Extreme Low Income line, by sex, age, employment status and geographical location (urban/rural)
1.2 By 2030, reduce at least by half (2015 Baseline) the proportion of men, women and children of all ages living in poverty in all its dimensions according to national definitions measured from 2015.	1.2.1 Proportion of population living below the national poverty line, by sex and age
	1.2.2 Proportion of men, women and children of all ages living in poverty in all its dimensions according to national definitions
1.3 Implement nationally appropriate social protection systems and measures for all, including floors, and by 2030 achieve 50% coverage of the poor and the vulnerable, identified without coverage in 2015.	1.3.1 Proportion of population covered by social protection floors/systems, by sex, distinguishing children, unemployed persons, older persons, persons with disabilities, pregnant women, newborns, work-injury victims and the poor and the vulnerable

D. B. Abraham (✉)
Rice University, Houston, TX, USA
e-mail: DAbraham@Rice.edu

D. B. Abraham, S. D. Iyer (eds.), *Promoting the Sustainable Development Goals in North American Cities*, Sustainable Development Goals Series, https://doi.org/10.1007/978-3-030-59173-1_11

1.5 By 2030, build the resilience of the poor and those in vulnerable situations and reduce their exposure and vulnerability to climate-related extreme events and other economic, social and environmental shocks and disasters.	1.5.1 Number of deaths, missing persons and persons affected by disaster per 100,000 people
	1.5.2 Direct disaster economic loss in relation to gross domestic product (GDP) for the area being measured
	1.5.3 Number of states with state level and local government disaster risk reduction strategies
1.a Ensure significant mobilization of resources from a variety of sources, including through enhanced development cooperation, in order to provide adequate and predictable means, to implement programs and policies to end poverty in all its dimensions	1.a.1 Proportion of resources allocated by the government directly to poverty reduction programs
	1.a.2 Proportion of total government spending on essential services (education, health and social protection)
1.b Create sound policy frameworks at the national, regional and international levels, based on pro-poor and gender-sensitive development strategies, to support accelerated investment in poverty eradication actions.	1.b.1 Proportion of government recurrent and capital spending to sectors that disproportionately benefit women, the poor and vulnerable groups

End hunger, achieve food security and improved nutrition and promote sustainable agriculture

2.1 By 2030, end hunger and ensure access by all people, in particular the poor and people in vulnerable situations, including infants, to safe, nutritious and sufficient food all year round.	2.1.1 Prevalence of obesity
	2.1.2 Prevalence of food insecurity in the population, based on percentage of the population living in food deserts (Food deserts are places more than 0.5 miles from a large grocery store selling fresh fruits and vegetables. Large grocery stores are typically stores with more than 50 employees and earning more than $2 million per year.)
2.4 By 2030, ensure sustainable food production systems and implement resilient agricultural practices that increase productivity and production, that help maintain ecosystems, that strengthen capacity for adaptation to climate change, extreme weather, drought, flooding and other disasters and that progressively improve land and soil quality.	2.4.1 Proportion of agricultural area under productive and sustainable agriculture

Ensure healthy lives and promote well-being for all at all ages

3.2 By 2030, end preventable deaths of newborns and children under 5 years of age, with all countries aiming to reduce under-5 mortality to at least as low as 25 per 1000 live births.	3.2.1 Under-5 mortality rate
	3.2.2 Low birth weight rate

3.3 By 2030, end the epidemics of AIDS, tuberculosis, malaria and neglected tropical diseases and combat hepatitis, water-borne diseases and other communicable diseases.	3.3.1 Number of new HIV infections per 1000 uninfected population, by sex, age and key populations
	3.3.2 Tuberculosis incidence per 1000 population
	3.3.3 Malaria incidence per 1000 population
	3.3.4 Hepatitis B incidence per 100,000 population
	3.3.5 Number of people requiring interventions against neglected tropical diseases
3.4 By 2030, reduce by one third premature mortality from non-communicable diseases through prevention and treatment and promote mental health and well-being.	3.4.1 Mortality rate attributed to cardiovascular disease, cancer, diabetes or chronic respiratory disease
	3.4.2 Suicide mortality rate
3.6 By 2020, halve the number of deaths and injuries from road traffic accidents.	3.6.1 Death rate due to road traffic injuries
3.8 Achieve universal health coverage, including financial risk protection, access to quality essential health-care services and access to safe, effective, quality and affordable essential medicines and vaccines for all.	3.8.1 Coverage of essential health services (defined as the average coverage of essential services based on tracer interventions that include reproductive, maternal, newborn and child health, infectious diseases, non-communicable diseases and service capacity and access, among the general and the most disadvantaged population)
	3.8.2 Number of people covered by health insurance or a public health system per 1000 population
3.9 By 2030, substantially reduce the number of deaths and illnesses from hazardous chemicals and air, water and soil pollution and contamination.	3.9.1 Asthma prevalence among adults >18 years
3.a Strengthen the implementation of the World Health Organization Framework Convention on Tobacco Control in all countries, as appropriate.	3.a.1 Age-standardized prevalence of current tobacco use among persons aged 18 years and older
3.c Substantially increase health financing and the recruitment, development, training and retention of the health workforce.	3.c.1 Health worker density and distribution

Ensure inclusive and equitable quality education and promote lifelong learning opportunities for all

4.1 By 2030, ensure that all girls and boys complete free, equitable and quality primary and secondary education leading to relevant and effective learning outcomes.	4.1.1 Proportion of children and young people: (a) in grades 2/3; (b) at the end of primary; and (c) at the end of lower secondary achieving at least a minimum proficiency level in (i) reading and (ii) mathematics, by sex
	4.1.2 High School Drop Out Rate

4.2 By 2030, ensure that all girls and boys have access to quality early childhood development, care and pre-primary education so that they are ready for primary education.	4.2.2 Participation rate in organized learning (one year before the official primary entry age), by sex
4.4 By 2030, substantially increase the number of youth and adults who have relevant skills, including technical and vocational skills, for employment, decent jobs and entrepreneurship.	4.4.1 Proportion of adults with at least a post-secondary degree
4.5 By 2030, eliminate gender disparities in education and ensure equal access to all levels of education and vocational training for the vulnerable, including persons with disabilities, indigenous peoples and children in vulnerable situations.	4.5.1 Parity indices (female/male, rural/urban, bottom/top wealth quintile and others such as disability status, indigenous peoples and conflict-affected, as data become available) for all education indicators on this list that can be disaggregated
4.7 By 2030, ensure that all learners acquire the knowledge and skills needed to promote sustainable development, including, among others, through education for sustainable development and sustainable lifestyles, human rights, gender equality, promotion of a culture of peace and non-violence, global citizenship and appreciation of cultural diversity and of culture's contribution to sustainable development.	4.7.1 Extent to which (i) global citizenship education and (ii) education for sustainable development, including gender equality and human rights, are mainstreamed at all levels in (a) national education policies, (b) curricula, (c) teacher education and (d) student assessment

Achieve gender equality and empower all women and girls

5.2 Eliminate all forms of violence against all women and girls in the public and private spheres, including trafficking and sexual and other types of exploitation.	5.2.2 Number of rape cases
5.5 Ensure women's full and effective participation and equal opportunities for leadership at all levels of decision-making in political, economic and public life.	5.5.2 Proportion of women-owned businesses 5.2.3 Gender wage gap

Ensure availability and sustainable management of water and sanitation for all

6.4 By 2030, substantially increase water-use efficiency across all sectors and ensure sustainable withdrawals and supply of freshwater to address water scarcity and substantially reduce the number of people suffering from water scarcity.	6.4.1 Water consumption 6.4.2 Level of water stress: freshwater withdrawal as a proportion of available freshwater resources

Ensure access to affordable, reliable, sustainable and modern energy for all

7.2 By 2030, increase substantially the share of renewable energy in the global energy mix.	7.2.1 Renewable energy share in the total final energy consumption
7.3 By 2030, double the global rate of improvement in energy efficiency.	

Promote sustained, inclusive and sustainable economic growth, full and productive employment and decent work for all

8.1 Sustain per capita economic growth in accordance with national circumstances and, in particular, at least 7% gross domestic product growth per annum in the least developed countries.	8.1.1 Annual growth rate of real GDP per capita
8.2 Achieve higher levels of economic productivity through diversification, technological upgrading and innovation, including through a focus on high-value added and labor-intensive sectors.	8.2.1 Annual growth rate of real GDP per employed person
8.5 By 2030, achieve full and productive employment and decent work for all women and men, including for young people and persons with disabilities, and equal pay for work of equal value.	8.5.1 Average hourly earnings of female and male employees, by occupation, age and persons with disabilities
	8.5.2 Unemployment rate, by sex, age and persons with disabilities
8.6 By 2020, substantially reduce the proportion of youth not in employment, education or training.	8.6.1 Proportion of youth (aged 15–24 years) not in education, employment or training

Build resilient infrastructure, promote inclusive and sustainable industrialization and foster innovation

9.c Significantly increase access to information and communications technology and strive to provide universal and affordable access to the Internet in least developed countries by 2020.	9.c.1 Proportion of population covered by a mobile network, by technology

Reduce inequality within and among countries

10.1 By 2030, progressively achieve and sustain income growth of the bottom 40% of the population at a rate higher than the national average.	10.1.1 Growth rates of household expenditure or income per capita among the bottom 40% of the population and the total population
	10.1.2 GINI Index of income inequality
	10.1.3 Racial segregation index

10.2 By 2030, empower and promote the social, economic and political inclusion of all, irrespective of age, sex, disability, race, ethnicity, origin, religion or economic or other status.	10.2.1 Proportion of people living below 50% of median income, by age, sex and persons with disabilities
10.6 Ensure enhanced representation and voice in decision-making in order to deliver more effective, credible, accountable and legitimate institutions.	10.6.1 Voting participation rate

Make cities and human settlements inclusive, safe, resilient and sustainable

11.1 By 2030, ensure access for all to adequate, safe and affordable housing and basic services and upgrade slums.	11.1.1 Proportion of urban population living in poor communities
	11.1.2 Housing and Transportation Affordability, number of persons spending more than 46% of income on housing and transportation costs
11.2 By 2030, provide access to safe, affordable, accessible and sustainable transport systems for all, improving road safety, notably by expanding public transport, with special attention to the needs of those in vulnerable situations, women, children, persons with disabilities and older persons.	11.2.1 Proportion of population that has convenient access to public transport, by sex, age and persons with disabilities
11.5 By 2030, significantly reduce the number of deaths and the number of people affected and substantially decrease the direct economic losses relative to global gross domestic product caused by disasters, including water-related disasters, with a focus on protecting the poor and people in vulnerable situations.	11.5.1 Number of deaths, missing persons and persons affected by disaster per 100,000 people[a]
11.6 By 2030, reduce the adverse per capita environmental impact of cities, including by paying special attention to air quality and municipal and other waste management.	11.6.2 Annual mean levels of fine particulate matter (e.g. PM2.5 and PM10) in cities (population weighted); annual measure for ozone
11.7 By 2030, provide universal access to safe, inclusive and accessible, green and public spaces, in particular for women and children, older persons and persons with disabilities.	11.7.1 Average share of the built-up area of cities that is open space for public use for all, by sex, age and persons with disabilities

Ensure sustainable consumption and production patterns

12.4 By 2020, achieve the environmentally sound management of chemicals and all wastes throughout their life cycle, in accordance with agreed international frameworks, and significantly reduce their release to air, water and soil in order to minimize their adverse impacts on human health and the environment.	12.4.3 Area of Brownfield or superfund sites

12.5 By 2030, substantially reduce waste generation through prevention, reduction, recycling and reuse.	12.5.1 National recycling rate, tons of material recycled
12.6 Encourage companies, especially large and transnational companies, to adopt sustainable practices and to integrate sustainability information into their reporting cycle.	12.6.1 Number of companies publishing sustainability reports
Take urgent action to combat climate change and its impacts[b]	
13.1 Strengthen resilience and adaptive capacity to climate-related hazards and natural disasters in all countries.	13.1.2 Number of deaths, missing persons and persons affected by disaster per 100,000 people[a]
13.2 Integrate climate change measures into national policies, strategies and planning.	13.2.1 Number of cities that have communicated the establishment or operationalization of an integrated policy/strategy/plan which increases their ability to adapt to the adverse impacts of climate change, and foster climate resilience and low greenhouse gas emissions development in a manner that does not threaten food production (including a national adaptation plan, nationally determined contribution, national communication, biennial update report or other)
13.3 Improve education, awareness-raising and human and institutional capacity on climate change mitigation, adaptation, impact reduction and early warning.	13.3.1 Number of cities that have integrated mitigation, adaptation, impact reduction and early warning into primary, secondary and tertiary curricula
	13.3.2 Number of cities that have communicated the strengthening of institutional, systemic and individual capacity-building to implement adaptation, mitigation and technology transfer, and development actions
13.b Promote mechanisms for raising capacity for effective climate change-related planning and management in least developed countries and small island developing States, including focusing on women, youth and local and marginalized communities.	13.b.1 Number of least developed countries and small island developing States that are receiving specialized support, and amount of support, including finance, technology and capacity-building, for mechanisms for raising capacities for effective climate change-related planning and management, including focusing on women, youth and local and marginalized communities – N/A
Protect, restore and promote sustainable use of terrestrial ecosystems, sustainably manage forests, combat desertification, and halt and reverse land degradation and halt biodiversity loss	

15.1 By 2020, ensure the conservation, restoration and sustainable use of terrestrial and inland freshwater ecosystems and their services, in particular forests, wetlands, mountains and drylands, in line with obligations under international agreements.	15.1.1 Forest area as a proportion of total land area
	15.1.2 Undeveloped Open Space as a proportion of total land area
	15.1.3 Proportion of important sites for terrestrial and freshwater biodiversity that are covered by protected areas, by ecosystem type
Promote peaceful and inclusive societies for sustainable development, provide access to justice for all and build effective, accountable and inclusive institutions at all levels	
16.1 Significantly reduce all forms of violence and related death rates everywhere.	16.1.1 Number of victims of intentional homicide per 100,000 population, by sex and age
	16.1.2 Conflict-related deaths per 100,000 population, by sex, age and cause
	16.1.3 Proportion of population subjected to physical, psychological or sexual violence in the previous 12 months
16.2 End abuse and rape.	16.2.3 Proportion of young women and men aged 18–29 years who experienced sexual violence by age 18
16.6 Develop effective, accountable and transparent institutions at all levels.	16.6.1 Primary government expenditures as a proportion of original approved budget, by sector (or by budget codes or similar)
16.7 Ensure responsive, inclusive, participatory and representative decision-making at all levels.	16.7.1 Proportions of positions (by sex, age, persons with disabilities and population groups) in public institutions (national and local legislatures, public service and judiciary) compared to national distributions
Strengthen the means of implementation and revitalize the Global Partnership for Sustainable Development	
Finance	
17.1 Strengthen domestic resource mobilization, including through international support to developing countries, to improve domestic capacity for tax and other revenue collection.	17.1.1 Total government revenue as a proportion of GDP, by source
	17.1.2 Proportion of domestic budget funded by domestic taxes
Technology	

17.6 Enhance North-South, South-South and triangular regional and international cooperation on and access to science, technology and innovation and enhance knowledge-sharing on mutually agreed terms, including through improved coordination among existing mechanisms, in particular at the United Nations level, and through a global technology facilitation mechanism.	17.6.2 Fixed Internet broadband subscriptions per 100 inhabitants
Systemic issues	
Policy and institutional coherence	
17.14 Enhance policy coherence for sustainable development	17.14.1 Number of countries with mechanisms in place to enhance policy coherence of sustainable development
Multi-stakeholder partnerships	
17.16 Enhance the Global Partnership for Sustainable Development, complemented by multi-stakeholder partnerships that mobilize and share knowledge, expertise, technology and financial resources, to support the achievement of the Sustainable Development Goals in all countries, in particular developing countries.	17.16.1 Number of countries reporting progress in multi-stakeholder development effectiveness monitoring frameworks that support the achievement of the sustainable development goals
Data, monitoring and accountability	
17.18 By 2020, enhance capacity-building support to developing countries, including for least developed countries and small island developing States, to increase significantly the availability of high-quality, timely and reliable data disaggregated by income, gender, age, race, ethnicity, migratory status, disability, geographic location and other characteristics relevant in national contexts.	17.18.1 Proportion of sustainable development indicators produced at the local level with full disaggregation when relevant to the target, in accordance with the Fundamental Principles of Official Statistics
	17.18.3 Number of cities with a local statistical plan that is fully funded and under implementation, by source of funding

References

Espey J, Dahmm H, Manderino L (2018) Leaving no U.S. city behind: The 2018 U.S. cities SDGs index. Sustainable Development Solutions Network, New York

Schmidt-Traub G, Kroll C, Teksoz K, Durand-Delacre D, Sachs J (2017) National baselines for the sustainable development goals assessed in the SDG index and dashboards. Nat Geosci 10:547–556

David B. Abraham, PhD (co-editor), is a scholar of Environmental Planning and Policy, specializing in Sustainability Planning and Performance Measurement. His research interests focus on developing sustainability strategies for the enhancement of living places. His research and professional experience includes food insecurity planning; health and well-being impacts from the built environment; urban forestry sustainability strategies; resiliency and revitalization community planning; and community performance metrics. His work has been featured at workshops, keynote agendas and conference proceedings throughout the USA and internationally in Ireland, Barbados, Brazil, Canada and Portugal. Dr. Abraham is a Research Scientist and Lecturer at Rice University.

Index

A
Action Program on Social Entrepreneurship, 100
Agriculture, 73
American Housing Survey (AHS), 111, 113
Applied Survey Research (ASR), 67, 68, 70

B
Baltimore Community Foundation (BCF), 11
Baltimore Neighborhood Indicators Alliance (BNIA), 8, 9, 13
Baltimore Neighborhood Indicators Alliance at the Jacob France Institute (BNIA-JFI)
  cities initiative, USA, 8
  SDGs, 19
    agency-promulgated rules and regulations, 18
    Data day, 14, 16
    generalizations, 18
    mapping, 17
    peace and justice, 16
    poverty/health, 16
    targets, 13
  stakeholders
    core project team, 9
    listening-to-the-listening, 10
    SDG-ET/working team, 9, 10
    sustainability plan, 10, 11
  staketaking, Sustainability Plan/initiative goals/target, 11, 12
  USA-SCI, 7, 8
Baltimore Office of Sustainability (BOS), 11
Brookings Metropolitan Policy, 127

C
California-based social research organization, 67
Child poverty, 89, 90
Children and Youth Well-being Spotlight (CYWB), 82
City-led programs, 100
City of Maple Ridge (CMR), 56–58
City stakeholders, 93, 95
Civic engagement, 79
Communities Without Boundaries International (CWBI), 9
Community assessment, 83
Community Assessment Project (CAP), 67
  collaborative leadership, 69
  culture of accountability, 70
  data-informed accountability, 69
  diverse sectors, 69
  organizations and institutions, 68
  plans and processes, 68
  Santa Cruz County, 67–69
Community Capital Framework (CCF), 54
Community Capital Tool (CCT)
  case studies context
    CMR, 56
    DNV, 56
    research methodology, 57, 58, 60
  complimentary instruments, 64
  developing sustainable communities, 53
  generalizations, 63
  goals/targets/indicators mapping, 59–61
  interview findings, 62, 63
  monitoring sustainable communities, 54, 55
  sustainable development, 51, 52
Community connectedness, 72
Community improvement cycle, 69
Community Indicator Consortium (CIC) session, 70
Community indicator systems (CISs)
  definition, 101
  new Tracking-Progress CIS platform, 104
  online, 98, 103
  online platforms, 97, 98
  Peg's mission, 102
  for SDG localization, 101
  SDG implementation, 97
  shared measurement system, 101
  web-based, 101
Community involvement, 84
Community well-being, 70
  bottom-up approach, 106
  in Canada, 99
  CIS use, 98
  community sustainability, 101
  indicator, 99
  Indigenous communities, 99
  in North America, 98
  in Winnipeg, 102

D. B. Abraham, S. D. Iyer (eds.), *Promoting the Sustainable Development Goals in North American Cities*, Sustainable Development Goals Series, https://doi.org/10.1007/978-3-030-59173-1

Complementary National Indicators, 127
Composition indicators
  housing quality index, 115
  value ratio index, 115
Continuous community improvement, 72
Culture of Health Alumni network, 70

D
Data availability, 91, 92
District of North Vancouver (DNV), 56, 58

E
Early childhood education (ECE), 74
Economic aspects, 2
English Learners (ELs), 75
Experimental Housing Allowance Program (EHAP), 112

G
Gender wage gap, 41

H
Health-care program insuring children, 70
Health insurance coverage, 76
Healthful diets, 77
Homelessness, 78
Housing and Urban Development (HUD), 111
Houston region
  development performance
    affordable/clean energy, 42
    clean water/sanitation, 42
    climate action, 46
    decent work/economic growth, 42–43
    gender equality, 41
    good health/well being, 39–40
    industry, innovation, and infrastructure, 43
    inequalities, reduced, 44
    life of land, 46–47
    metros, 35–37
    no poverty, 38
    peace/justice/strong institution, 47
    quality education, 40–41
    steady-state economy, 47
    sustainable cities/communities, 44–45
    Zero hunger, 39
  performance, 48, 49
  port, 34, 35
  sustainability efforts, 31, 32
  sustainability/state of growth and development, 32–35

I
Individual well-being, 70
Inter-Agency and Expert Group on Sustainable Development Goal Indicators (IAEG-SDG), 94, 95

L
LEHD Origin-Destination Employment Statistics (LODES), 95
Liquid asset poverty, 94
Local Data Action (LDA) project, 93

M
Massachusetts Institute of Technology (MIT), 17
Maternal mortality data, 92
Metropolitan statistical area (MSA), 88, 89, 91, 92
Millennium Development Goals (MDGs), 2, 23, 52, 100

N
National Neighborhood Indicators Partnership (NNIP), 8
Natural environment, 80, 81
Neighborhood University Community Council (NUCC), 113

O
Official Community Plan (OCP), 56
One Baltimore, 7
Organisation for Economic Co-operation and Development (OECD), 71, 127

P
Peg (online CIS)
  indicator framework, 102
  indicators, 105
  mission, 102
  outreach and education strategy, 105
  Peg 1.0 software, 102, 103
  Peg 2.0, 104–105
  preliminary work and development, 99
  resources, 102
  technology and updating data, 102
  2-year process, 98
  Winnipeg's SDG localization process, 105
Political engagement, 91
Poor air quality, 81
Pregnancy, 76
Public Safety, 79

R
Racial inequalities, 89
Renewable energy, 92
Results-based accountability (RBA), 67, 72
Return on investment (ROI), 131

S
SDG implementation, 87
SDG Index, 88
SDG localization
  CISs, 97, 101
  communities, 98

community well-being measurement, 99
data dashboard process, 100
framework, 100
global effort, 100
implementation toolkits, 101
in North America, 105
regional and local governments, 100
regional programs, 100
sub-national/local level, 100
UN SDG indicator framework, 98
VNRs, 100
Winnipeg's SDG localization process, 105
SDG monitoring
benefits and limitations, 88
bottom-up approach, 88
data monitoring mechanisms, 95
exercises, 94
LDA project, 93
local SDG monitoring efforts, 93–94
SDSN, 93
SDGs Executive Team (SDG-ET), 9
Social connectedness, 67, 70, 71, 82
Social Determinants of Health (SDOH), 73–75, 78, 82–84
Social-emotional learning (SEL), 129
Social inequalities, 89
Social Initiative Fund (SIF), 131
Sprawl, 32
Substandard housing, sustainability indicator
collecting data, 110
comparing two composite indicators, 122, 123
measuring inadequate housing, 110–113
methodology, 113, 114, 124
quality, 110
results/analysis, 115–118
statistical tests, 110
testing significance, difference age of units, 119–122
Sundance Family Foundation, 5
Sustainable development
comprehensive and inclusive framework, 1
five Ps, 2
Houston stakeholders, 4
inclusivity, 2
integration, 2
ITC and data availability, 2
local achievements, 3
principles in development strategies, 2
stakeholders, 3
universality, 2
Sustainable development goals (SDGs), 7, 31, 51, 109, 127, 129
agriculture, 73, 148
aligning, 24, 25
indicator alignment, 26–28
policy alignment, 25, 26
ambitious goals, 21
availability/sustainable management, 150
bottom-up approach, 97
CAP, 67
challenges, 1
cities/human settlements, 152
city planning, 1
climate action, 79
climate change, 1, 153
consumption/production patterns, 152
critical issues, 2
CYWB, 82
decent work and economic growth, 73
deteriorating infrastructure, 1
ECE programs, 74
economic development, 1
economic growth/full/productive employment, 151
environmental development, 1
equitable quality education, 149
experience, Los Angeles, 28
gender equality/empower, 150
global partnership, 154, 155
good health and well-being, 75
healthy lives, all age, 148
implementation, 22, 98
job loss, 1
key factors, 23, 24
life below the water, 79
life on land, 79
Los Angeles, 22, 23
maximize value, 29, 30
national mobilization and local engagement, 4
no poverty, 73
peace, justice and strong institutions, 79
quality education, 74
reduce inequality, 151
reduced inequalities, sustainable cities and communities, 78
resilient infrastructure, 151
social development, 1
social exclusion, 1
societies, 154
sustainable development, 73
targets and indicators, 2
targets and metrics, 21
terrestrial ecosystem, 153
tourism, 73
zero hunger, 73
Sustainable Development Solutions Network (SDSN), 7, 17, 58
global goals, 87
LDA project, 93
National Index for US Cities, 88, 89
partners, 93
technical exercise, 87
USA-SCI, 92

T

Telephone Consumer Protection Act (TCPA), 72
Together for Youth, 70, 82
Tourism, 73
Tracking-Progress platform, 102–106
Transportation, 34

U
University of Baltimore (UB), 9
University of Maryland (UMD), 9
Unsafe neighborhoods, 80
USA Sustainable Cities Initiative (USA-SCI), 58, 92

V
Voluntary National Reviews (VNRs), 24, 29, 97, 100

W
Web-based CISs, 101
Work Readiness assessments, 134
Works Progress Administration (WPA), 111
World Council on City Data (WCCD), 27

Y
Youth Development and Housing, 4
Youth Homeless Demonstration Project, 79
Youth Social Entrepreneurship (YSE), 129
  common indicators, uncommon youth serving nonprofits, 134
  common set indicators, 133
  community and cultural engagement, 130
  defining, 130
  developing common instruments, 132
  entrepreneurship, 130
  nonprofits, Amercian cities
    key overall findings, 139, 142
    UN SDGs, 142, 143
  outcomes
    age/acquisition skills, 138
    age/gender, 138
    length of involvement, 137
    social-economic development, 138, 139
  reliability, 134
  social-emotional development, 130
  workforce development system, 143, 144
  workforce readiness, 130
  world Spark agency, 132, 133
Youth Violence Prevention Task Force (YVPTF), 70, 81, 82